from
BLUE MOONS
to
BLACK HOLES

"Melanie Melton Knocke has written an excellent book that should find its way to the bookshelf of anyone interested in astronomy. Anyone interested in astronomy will find the 'Quick Facts' section a valuable resource. She has collected in one place pertinent information about the planets, their moons, astronomical events, stars, and so on. Even more valuable is the 'Brief History of Lunar and Planetary Exploration' section, which collects information about every mission to the Moon and planets in one place. Finally, there is a single 'go-to' book for this type of information."

—Greg Novacek, Director,
Lake Afton Public Observatory, Wichita, Kansas

"Melanie Melton Knocke's new book, *From Blue Moons to Black Holes*, is an informative and friendly introduction to the universe. The book provides quick, accurate answers to common questions about space exploration, the solar system, and the universe. Each answer also contains a detailed explanation that allows the reader to explore the topic in greater depth. I highly recommend *From Blue Moons to Black Holes* for the budding young astronomer, the seasoned amateur, and all who are curious about the universe."

—Scott Kardel, Public Affairs Coordinator,
California Institute of Technology's Palomar Observatory

"Questions—simple questions—about everything up there, is what this book is about. This book is a treasure of clear answers to common questions. Find a handy place for it on your bookshelf, because you will refer to it often!"

—David H. Levy, author of *Cosmic Discoveries* and *Starry Night*

from BLUE MOONS to BLACK HOLES

A Basic Guide to Astronomy, Outer Space, and Space Exploration

Melanie MELTON KNOCKE

Prometheus Books

59 John Glenn Drive
Amherst, New York 14228-2197

Published 2005 by Prometheus Books

Inquiries should be addressed to
Prometheus Books
59 John Glenn Drive
Amherst, New York 14228–2197
VOICE: 716–691–0133, ext. 207
FAX: 716–564–2711
WWW.PROMETHEUSBOOKS.COM

09 08 07 06 05   5 4 3 2 1

Library of Congress Cataloging-in-Publication Data

Knocke, Melanie Melton.
    From blue moons to black holes : a basic guide to astronomy, outer space, and space exploration / Melanie Melton Knocke.
        p. cm.
    Includes bibliographical references and index.
    ISBN 1–59102–288–6 (alk. paper)
    1. Astronomy—Popular works. 2. Solar system—Popular works. 3. Planets—Exploration—Popular works. I. Title.

QB44.3.K66 2005
520—dc22

                                                                                    2005001754

Printed in the United States of America on acid-free paper

To *Phil*

CONTENTS

PART 3: A BRIEF HISTORY OF LUNAR AND PLANETARY EXPLORATION 225

ACKNOWLEDGMENTS

This book would not have been possible without the help of many people. Listed below are just a few. I hope I got them all, and sorry if I missed someone.

Philip Knocke, Jet Propulsion Laboratory—fact checker, proofreader extraordinaire, babysitter, cook, and much more. Helen Melton and Corkie Martin—proofreaders. Bob Melton, Jack and Eunice Knocke—babysitters. Greg Novacek and Jim Fullerton, Lake Afton Public Observatory—for images and advice. Tim Rodriquez, Lowell Observatory—for helping me understand the Kuiper Belt. Dave Schleicher, Lowell Observatory—for the great Hale-Bopp image. Ken Baun, Meade Instruments—for putting me in touch with Jack Newton. Jack Newton, Arizona Sky Village—for the great images. Fred Espenak, Goddard Space Flight Center—for the use of his eclipse predictions. Lou Friedman, Charlene Anderson, Monica Lopez, Donna Stevens, and Pete Hernandez, The Planetary Society—for things too numerous to mention. Louise Ketz, Ketz Literary Agency—for being on my side. Linda Regan, Prometheus Books—for believing in the book. KOA Kampgrounds—for having modem connections at almost all your campgrounds. All the engineers and scientists who work to help us better understand the world in which we live. And Charlie—thanks for sharing your mommy with this book.

INTRODUCTION

*F*rom *Blue Moons to Black Holes* is written specifically for those who have an interest in astronomy and space but little time to explore the amazing world of exploding stars, distant galaxies, rovers on other planets, and more. The book consists of three sections: Questions and Answers, Quick Facts, and a Brief History of Lunar and Planetary Exploration.

In Questions and Answers, the reader finds simple and easy to understand answers to the most common questions people have regarding astronomy, outer space, and space exploration. Many of the questions come directly from my work with the general public as Assistant Director of Public Programs at Lowell Observatory in Flagstaff, Arizona; Education Program Specialist at Mount Wilson Observatory in Pasadena, California; and Assistant Director of the Lake Afton Public Observatory in Wichita, Kansas. Sample questions include: "What is a blue moon?" "Could you travel through a black hole?" "Is the North Star the brightest star in the sky?" and "Is Pluto really a planet?" To aid the reader, the answer to each question is complete in itself.

Quick Facts offers the reader an easy way to look up interesting statistics about the Moon and planets, bright stars, constellations, and more. This section also includes guides to upcoming meteor showers as well as lunar and solar eclipses.

The Brief History of Lunar and Planetary Exploration includes a chronological listing of every mission that has been launched to the Moon and planets to date. By listing both the successes and the failures, readers gain a better understanding of just how difficult it is to travel beyond our own planet. In addition, interesting facts emerge, such as the mission that sent turtles and worms around the Moon, or the debris from a mission failure that was initially mistaken for a Russian ICBM missile attack.

Our universe is a fascinating place full of exotic entities like black holes and blue moons, white dwarfs and red giants. And it's out there for anyone who takes the time to look up! As this book will show, you don't need a degree in physics to explore the vast reaches of outer space. All you need is curiosity and a little imagination.

Enjoy!

PART 1

QUESTIONS AND ANSWERS

ABOUT THE SOLAR SYSTEM

Our Sun.
Image credit: Royal Swedish Academy of Sciences

What exactly is the solar system?
Everything within the Sun's influence.

The Latin name for our Sun is *Sol*. Solar system, or the Sun's system, is the name given to those things that fall under the influence of the Sun's gravity and energy output.

Objects within our solar system include our star (the Sun), the nine planets and their many moons and ring systems, asteroids, comets, and meteors. Our solar system also includes large numbers of planetesimals (small icy bodies) within the distant Kuiper Belt and Oort Cloud. Even tiny dust and ice particles floating in interplanetary space are considered to be part of our solar system.

How big is the solar system?
At least eight billion miles (thirteen billion kilometers) and counting.[1]

For many years, it was believed that Pluto marked the edge of our solar system at a distance of four billion miles (six billion kilometers). However, recent discoveries have shown that it is at least double that size. And, as equipment and observing techniques improve, we may find out that our solar system is even larger.

In the outer reaches of the solar system, the Sun's influence is mainly seen in the form of a weak gravitational and magnetic field as well as a slight solar wind (a stream of charged particles flowing away from the Sun). Scientists are using two venerable spacecraft to try and detect just exactly where the solar wind fades out and interstellar space takes over. The spacecraft, *Voyager 1* and *Voyager 2*, are currently on their way out of the solar system, having long ago completed their primary mission of studying Jupiter, Saturn, Uranus, and Neptune. As the spacecraft head toward oblivion, scientists are receiving information about the space through which they are flying.

By July 2005, *Voyager 1* (the more distant of the two) will be approximately 8.8 billion miles (14.3 billion kilometers) from the Sun.[2] Although the spacecraft is still detecting evidence of solar wind at this distance, the wind's measured composition has changed slightly, which suggests that the Sun's influence is starting to fade.[3] The power supplies on the two spacecraft are expected to last until the year 2020.[4] Perhaps by then scientists will have seen the last of the solar wind.

In the meantime, ground-based telescopes have detected many small icy bodies beyond Pluto, including a small reddish object dubbed Sedna (after the Inuit goddess of sea creatures). First detected at a distance of eight billion miles (thirteen billion kilometers), this is the most distant solar system object currently known. Sedna has a highly elliptical orbit. At its closest point, this small body is still twice as far from the Sun as Pluto. At its most distant point, Sedna is a whopping 84 billion miles (130 billion kilometers)[5] from the Sun—much too distant to be seen from Earth. While Sedna appears to lie beyond much of the Sun's influence, it is still attached to our star by a weak thread of gravity.

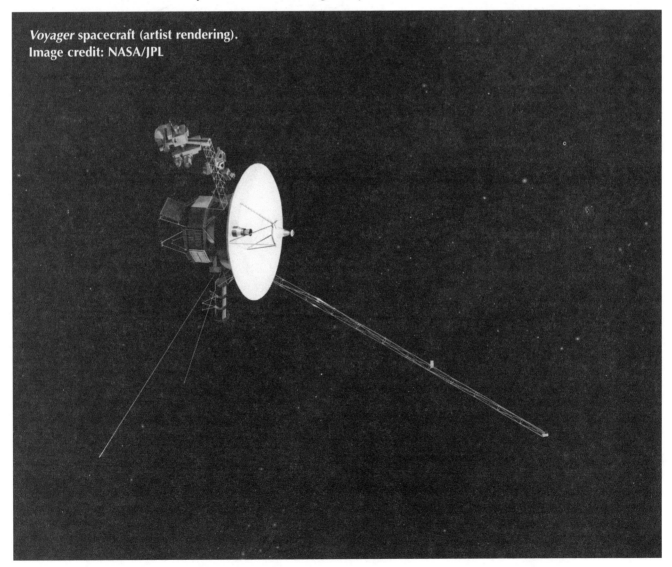

Voyager **spacecraft (artist rendering).**
Image credit: NASA/JPL

Astronomers are continuing the search for objects at or near the outer fringes of our solar system. It is hoped that time and new discoveries will give us a definitive answer as to the true size of our solar system. For now, you can safely say that it has a radius of at least eight billion miles (thirteen billion kilometers).

How did the solar system form?
From the gravitational collapse of a large rotating cloud of gas and dust.

Since scientists have yet to build a time machine that would allow them to witness the formation of our own solar system, they don't know exactly how the Sun and planets formed. However, by observing other star systems in various stages of evolution, they think they have a pretty good idea of what occurred.

The current theory of solar system evolution has our Sun and planets forming from a rotating disk of gas and dust.[6] Originally this disk was a large stellar cloud known as a diffuse nebula. At some point, the distribution of matter within this cloud became uneven. Where there was a slightly larger concentration of gas and dust (a clump), the combined gravity of the clump began attracting other nearby gas and dust to it.

As more matter was attracted to the original concentration, its gravity increased as well. In addition, as the new matter was pulled into the clump, it collided with the original material, causing the entire clump to begin to rotate. With the increased mass, even more material was attracted to the clump. Soon the situation was out of control, with the gravity and mass increasing at a dramatic rate.

The gravity of this large concentration (now a protostar) attracted so much matter from the surrounding nebula that it couldn't all be pulled in at once. Instead, material was first drawn into a disk surrounding the protostar. This flattened disk stretched for billions of miles in every direction and probably resembled a whirlpool with matter spiraling into the protostar at the center.

As the mass and gravity of the proto-star continued to increase, so did the pressure and temperature in its core. Finally, conditions in the core reached a point where hydrogen atoms were being squeezed so tightly together that they began to fuse. When two atoms are forced to fuse together, the process is called nuclear fusion. The result of hydrogen nuclear fusion is the creation of a helium atom and the release of a tremendous amount of energy. With the beginning of nuclear fusion in its core, our Sun was born.

At this point, the energy released by the nuclear fusion reactions pushed outward from the Sun's interior and stopped its gravitational collapse. Matter that had been in the process of spiraling in toward the star was suddenly left out in the cold. The leftover material continued to orbit around the newly formed star. Eventually this material started to come together in larger and larger concentrations. Over a period of a few thousand years, matter within the disk coalesced into the planets we know today. In the region of space between Mars and Jupiter, material was unable to come together to form a large body because Jupiter's massive gravity kept disrupting the area. Instead, thousands of smaller bodies formed to create the asteroid belt.

Beyond Neptune the disk did not have a distinct edge, but rather thinned out gradually the farther it was from the Sun.[7] Beyond the point where Neptune formed, matter was so thinly distributed that it was difficult for it to amass into a large planet. Instead, localized areas of dense material came together to form small icy bodies, each a few hundred miles across. Thousands of these bodies formed in a region we know today as the Kuiper Belt.

The events described above would have taken at least a few million years to complete. So even if scientists ever invent a time machine, they would probably still have trouble witnessing the entire event. For now we'll have to rely on their theories that are based on scientific evidence.

How old is our solar system?
Approximately four and a half billion years old.

Determining the age of the solar system is not as easy as one might think. Trying to find out the precise age of a planet is extremely difficult, if not impossible. For example, the original surface of the Earth has been altered beyond recognition by the movement of its crust by plate tectonics, volcanism, earthquakes, weathering, and the like. Jupiter, Saturn, Uranus, and Neptune are made almost entirely of gas and have no surface to study, therefore offering even fewer clues about their age. Moons and asteroids offer a few hints, but even their surfaces have often been altered by collisions.

The only source in the solar system that offers astronomers a fairly reliable age is the Sun. The surface of the Sun changes from moment to moment, so it is of no use in this calculation. Instead, astronomers look deeper for their information. Within the Sun's core, hydrogen is being fused into helium at a tremendous rate. Through information about the Sun's mass, diameter, and composition, astronomers can calculate how fast these reactions are taking place and how much fuel the Sun has left to burn. They can also backtrack to determine when the nuclear reactions must have begun.

From all of this, they calculate that the Sun has been fusing hydrogen for approximately four and a half billion years. (For more information, see "About the Sun" beginning on page 27.)

How do astronomers measure distances to the planets?
They use a mathematical formula known as Kepler's third law and a few other modern techniques as well.

To determine the average distances to the planets (or comets, asteroids, or any other body in orbit around the Sun), astronomers use a nifty little equation known as Kepler's third law. All they need to know beforehand is how long the object in question takes to orbit the Sun (its period). Once they know the object's period, they simply plug that value into Kepler's equation and out comes its average distance.

(For those mathematically inclined individuals, Kepler's third law is written as $P^2 = a^3$ where P is the object's period in years and a is the object's semimajor axis in astronomical units—or the average of the object's greatest distance and minimum distance from the Sun).[8]

Kepler's third law has been around since the early 1600s when the German mathematician Johannes Kepler discovered the relationship between a planet's distance and orbital period during his careful examination of the planetary observations of the Danish astronomer Tycho Brahe.[9] This simple relationship allowed astronomers to accurately calculate the distance to all the known planets of the time—Mercury, Venus, Mars, Jupiter, and Saturn. The equation continued to work with the discovery of the more distant planets as well as the asteroids within the asteroid belt. Even today, when a new asteroid or object beyond Pluto is discovered, Kepler's third law is used to give astronomers its average distance.

The original version of Kepler's third law is accurate only for planets and other objects in

Kepler's third law can be used to calculate the distance to Saturn and the other planets.
Image credit: Hubble Heritage Team (NASA/AURA/STScI)
Acknowledgment: R. G. French (Wellesley College), J. Cuzzi (NASA/Ames),
L. Dones (SwRI), and J. Lissauer (NASA/Ames)

orbit around the Sun. A slightly modified version of this law can be used to determine the distance between any two bodies in orbit around each other, whether the pair includes a moon and a planet, a spacecraft and a planet, or two stars. (Kepler's first two laws explain the elliptical shape of a planet's orbit and its travel time around a star.)

In addition to Kepler's third law, astronomers also use modern techniques such as bouncing a laser beam off the Moon's surface or a radar signal off the clouds of Venus. Timing how long it takes these signals to return to Earth tells astronomers how far the signal had to travel and thus how far away these bodies are from our planet.

What is an astronomical unit (or AU)?
A unit of measure equal to 93,000,000 miles (150,000,000 kilometers), the average distance between Earth and the Sun.

The astronomical unit is a somewhat cleaner way to keep track of the distances of the planets. For example, instead of having to write Earth's distance as 93,000,000 miles (150,000,000 kilometers), one can simply say Earth is one astronomical unit, or 1 AU, from the Sun. Below is a list with the distances of the planets given in astronomical units:

Planet	Distance from the Sun (in AUs)
Mercury	0.39
Venus	0.72
Earth	1.0
Mars	1.5
Jupiter	5.2
Saturn	9.55
Uranus	19.2
Neptune	30.1
Pluto	39.4

Note: When using Kepler's third law of planetary motion to determine the average distance of an object from the Sun (as discussed in the previous question), the distances are given in astronomical units.

What lies beyond the orbit of Pluto?
Small bodies composed of ice and rock—
probably debris left over from the formation of our solar system.

Because of their small sizes and great distances from Earth, objects beyond Pluto are extremely difficult to observe. It has only been within the last few years that telescopes and cameras have become sensitive enough to detect these distant bodies. Since 1992, over eight hundred such objects have been discovered, with more being found each year.[10]

The classification of these objects has yet to be officially sorted out. Unofficially, most astronomers use the term Kuiper Belt object (KBO). Some call these objects Transneptunian objects (TNO). Still others refer to these distant objects as comets, asteroids, small planetesimals, or even "Plutinos." The classification of these objects will be resolved in time, as more information is gathered and we better understand the nature and makeup of these distant bodies.

What is the Kuiper Belt?
A region of space beginning near the orbit of Pluto
and stretching outward to 500 AUs.

The Kuiper Belt is home to thousands of small bodies believed to be leftover debris from the formation of our solar system. Many believe this debris to be potential comets waiting for a trip to the inner solar system. While the idea of this comet reservoir has been around since the early 1950s, the first object to be discovered within its predicted boundaries wasn't found until 1992.

The first official Kuiper Belt object (KBO) to be discovered, 1992 QB1, lies just beyond the orbit of Pluto and is thought to be no larger than 155 miles (250 kilometers) in diameter.[11] Since 1992, over eight hundred Kuiper Belt objects have been discovered, with more being found every year. Because of the great distances involved and their small sizes, not much is known about these objects. They are believed to be made up of ice and rock. Some have a reddish tint to them, while others are more neutral in color.

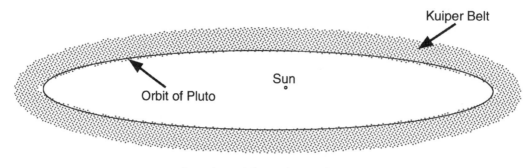

Location of the Kuiper Belt.

Currently, the largest KBO has been dubbed "2004 DW." Discovered in February 2004, 2004 DW is approximately 990 miles (1,600 kilometers) in diameter or half the size of Pluto.[12] It also orbits at about the same distance as our ninth planet. However, 2004 DW is not alone. There are at least two other KBOs (Quaoar and 2001 KX76) that are almost as large and have similar orbits.[13]

It was the discovery of objects with orbits similar to Pluto's that started the debate about Pluto's status in the solar system.[14] It now appears that there are several KBOs that either share Pluto's orbit or have an orbit very similar to that of our ninth planet. In addition, Pluto and the KBOs seem to share a similar composition of ice and rock. As a result, some astronomers consider Pluto to be a Kuiper Belt object instead of a planet. (See the question "Is Pluto really a planet?" on page 67 for more information.)

What are Transneptunian objects?
Small icy bodies that orbit beyond the planet Neptune—
also known as Kuiper Belt objects.

The term Transneptunian object (TNO) is another way to refer to any body orbiting beyond Neptune—including the planet Pluto. Astronomers who use this phrase are usually the ones who believe that Pluto should not be classified as a planet. Others use this phrase interchangeably with Kuiper Belt objects. However, this phrase appears to be fading in popularity as more and more astronomers adopt the concept of the Kuiper Belt.

What is the most distant known object in the solar system?
2003 VB12 (dubbed "Sedna").[15]

Discovered on November 14, 2003, this small world holds the record for being the farthest solar system object detected to date.[16]

Sedna's highly elliptical orbit brings it to within 8 billion miles (13 billion kilometers) of the Sun at its closest point. At its most distant point, the small body is a whopping 84 billion miles (130 billion kilometers) away. With this orbit, it takes Sedna 10,500 years to travel once around the Sun. For comparison, Pluto averages only 4 billion miles (6 billion kilometers) from the Sun and takes a mere 247.7 years to complete one orbit. In other words, our ninth planet could travel around the Sun forty-two times in the amount of time it takes Sedna to complete one orbit.

When the Spitzer Space Telescope observed Sedna, it was unable to detect any infrared radiation coming from the object. This lack of any detectable heat told astronomers a couple of things about this distant world. First, and probably the most obvious, it showed astronomers that Sedna is a very cold place. In fact, with a temperature of −400°F (−240°C), Sedna now holds the record for being the coldest object in the solar system.[17] The lack of detectable heat also suggested that Sedna is approximately one thousand miles (sixteen hundred kilometers) in size, or almost three-quarters the size of Pluto. Any smaller and it would be too faint to be seen from Earth at its current distance, any larger and there should be detectable infrared radiation coming from its surface. A diameter of one thousand miles makes Sedna the largest solar system object to be found since the discovery of Pluto.

Astronomers continue to study Sedna using both new and old techniques to draw out any information they can from the extremely faint light they are detecting from this distant world. Some astronomers believe that Sedna may belong to the Oort Cloud.

What is the Oort Cloud?
A distant sphere of material located almost 50,000 AUs from the Sun.

The Oort Cloud was first proposed by Dutch astronomer Jan Oort in 1950 as a way to explain the origins of medium- and long-period comets (comets that take longer than two hundred years to orbit the Sun).[18] At the time, there were some major theoretical problems that kept such comets from fitting nicely into any known solar system model. For one thing, these comets had extremely elliptical orbits with long periods—thousands of years in some cases— suggesting that they were originating in the outer reaches of the solar system. According to the current solar system model, it would not have been possible for anything to form at such a great distance from the Sun. Yet the sheer number of new comet discoveries, almost one a month, began to suggest an almost limitless supply.[19]

The orbits of many of these comets were also highly tilted, allowing them to enter the inner solar system from every direction—another fact contrary to the current solar system model.[20] Objects within our solar system formed within a flattened disk of gas and dust that was gravitationally attracted to a coalescing Sun. As a result, almost all solar system objects orbit within the same plane around the Sun, the ecliptic. Over time, the orbits of some objects have been tweaked by gravitational encounters, but even these bodies still orbit within an acceptable distance of the ecliptic.

By passing through the inner solar system from every direction, the medium- and long-period comets appeared to be breaking theories as well as the laws of gravitational attraction. However, after studying their orbits and compositions, Oort suggested that these comets were not breaking any laws after all. He believed that these comets did not form in their current orbits, but had originated somewhere between the orbits of Jupiter and Neptune.

Oort theorized that soon after these small icy bodies had formed, gravitational encounters with the larger planets flung many of them outward, both above and below the plane of the ecliptic. Though it is likely that some were flung completely out of our solar system, many remained bound to the Sun, but in much larger, more inclined orbits. Then an occasional encounter with a nearby star tweaked their orbits even further, until the reservoir resembled a sphere completely surrounding our solar system. Oort reasoned that a sphere-shaped reservoir, with the solar system in the center, would allow comets to pass through the inner solar system from any angle.

A recent analysis of the long-period comet Hale-Bopp bolsters this conclusion. When Hale-Bopp passed through the inner solar system in 1997, astronomers detected the presence of argon and the absence of neon in the comet's composition.[21] Since neon evaporates at −414°F (−248°C), the comet had to form where the temperature was warmer than −414°F. Because argon evaporates at −387°F (−233°C), and Hale-Bopp still seems to have plenty of argon, the comet could not have been exposed to temperatures higher than −387°F during its formation. This means the comet began its life in a region of space warmer than −414°F and cooler than −387°F, or somewhere between the orbits of Uranus and Neptune.

While it is believed that thousands of these icy bodies exist within this reservoir, their total mass is probably only equal to about three times the mass of Earth. When Sedna was first discovered, astronomers believed they had located the first member of the Oort Cloud. Now they aren't so sure. The current understanding of the Oort Cloud leaves little room for Sedna, whose maximum distance of 900 AUs places it much too close to the Sun for it to be an Oort Cloud member. However, some astronomers are now suggesting the existence of an inner Oort Cloud—one that exists between the Kuiper Belt and the original Oort Cloud.[22] Only time and more discoveries of Sedna-like objects will tell.

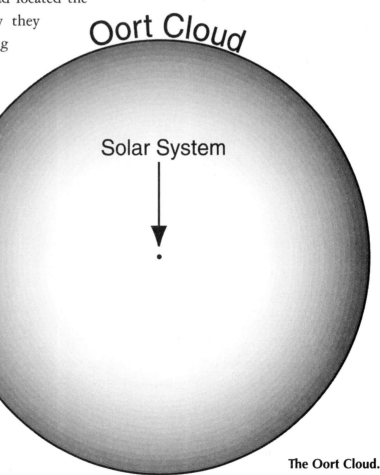

The Oort Cloud.

ABOUT THE SUN

What is the Sun?

What is sunlight?

How far away is the Sun?

How big is the Sun?

Is the Sun burning?

How hot is the Sun?

What is the solar wind?

What is a sunspot?

Are there always sunspots on the Sun?

What is a solar flare?

What is a coronal mass ejection?

What is the Sun made of?

If the Sun is made of gas, could you fall right through it?

Will our Sun ever burn out?

Will the Sun just burn out and go black?

What is a solar eclipse?

Why doesn't a solar eclipse happen every month—with each New Moon?

When is the next solar eclipse?

Why shouldn't you look directly at the Sun?

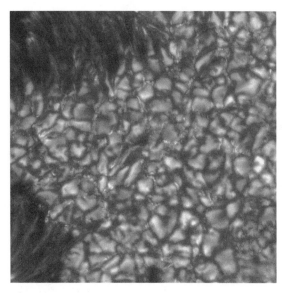

A close-up of the Sun's turbulent surface.
Image credit: Royal Swedish Academy of Sciences

What is the Sun?
The Sun is a star, a huge sphere of hydrogen and helium gas powered by nuclear fusion at its core.

Because the Sun looks so different from all of the other stars seen at night, many people don't realize that our Sun is also a star. The main difference between the nighttime stars and the Sun is not size or brightness, but distance. It is simply much closer to us.

For an earthly example: imagine that you are outside at night with two friends. Each friend has the same type of flashlight. One is standing three feet away from you and the other is standing twenty feet away. Now both of them shine their flashlights in your eyes. Which of the two flashlights would appear brighter? The one closest to you, of course. Is it really brighter? No, it is just closer.

Even though the Sun looks enormous in our skies, its size, temperature, and life expectancy are all considered average when compared to other stars. There are many stars larger than our Sun, as well as smaller. There are hotter and cooler stars as well. The wonderful thing about our Sun is that the very properties that make it an average star provide just the right conditions for life to exist on Earth. With a cooler or hotter star, it is likely that life would not have developed on our planet.

What is sunlight?
The visible portion of the Sun's radiation.

The Sun emits a tremendous amount of radiation at all wavelengths of the electromagnetic spectrum, including gamma rays, x-rays, ultraviolet radiation, visible light, infrared radiation, and radio waves. While Earth is bathed in all of this energy, our atmosphere blocks a great deal of the high-energy radiation (gamma rays, x-rays, and ultraviolet radiation) and a portion of the infrared as well. This is a good thing for all the sensitive-skinned creatures that live on this planet. Without the protection of the atmosphere, the high-energy radiation would fry us and the infrared radiation would really heat things up! One portion of the Sun's energy that our atmosphere does let through is the very portion to which the human eyes are most sensitive—visible light.

Sunlight is actually a combination of all the visible wavelengths of light. When sunlight is broken down into its individual wavelengths, as when it passes through a prism or a raindrop, we see all the colors of the rainbow: red, orange, yellow, green, blue, indigo, and violet.

How far away is the Sun?
On average, the Sun is 93 million miles (150 million kilometers) from Earth.

Because Earth's orbit is elliptical (egg-shaped), its distance from the Sun can vary by as much as three million miles, depending on where it is along its orbit. At its closest point, Earth is 91,350,000 miles (147,000,000 kilometers) from the Sun.[1] At its farthest point, Earth is 94,450,000 miles (152,100,000 kilometers) away.[2] The average distance between the two (93 million miles, or 150 million kilometers) is also known as an astronomical unit (AU).

Ninety-three million miles is a long ways away. From Earth, it would take a spacecraft traveling at a realistic cruising speed of 23,000 miles per hour (37,000 kilometers per hour) almost six months to reach the Sun. However, as distances in space go, it's really not that far. If you wanted to travel to the next closest star system, Alpha Centauri, which is almost 25 trillion miles (40 trillion kilometers) away, it would take the same spacecraft over one hundred thousand years to get there.[3] And that is our closest neighbor. Imagine if you wanted to explore the next galaxy.

So the Sun may seem as though it is really far away, but it is really just next door when compared to the other stars in our sky.

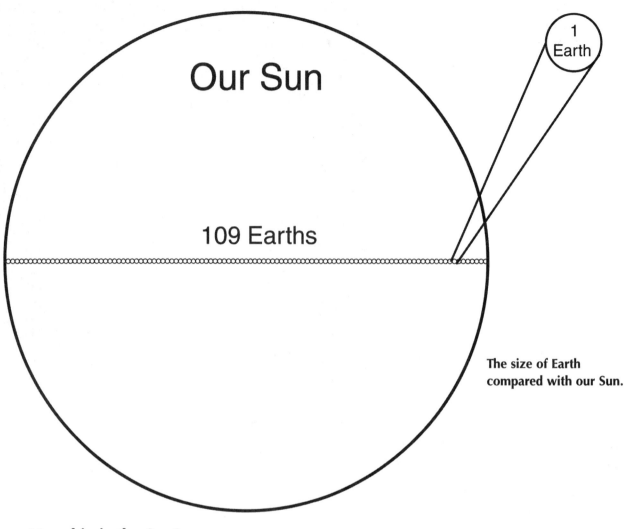

Our Sun

109 Earths

1 Earth

The size of Earth compared with our Sun.

How big is the Sun?
864,400 miles (1,392,000 kilometers) in diameter.[4]

Even though our Sun is considered an average star, it is by far the largest and most massive object in our solar system. It is so large, 109 Earths could be lined up across the Sun's disk.

Is the Sun burning?
Yes, but not in the same way as a fire in the fireplace.

Comparing the way the Sun is burning to a fire in the fireplace is like comparing a fireplace to a hydrogen bomb. The first will warm a few people with its heat; the other will level a large city. They are two totally different types of reactions.

A fire in the fireplace uses wood for fuel. Burning the wood to produce heat and light is a type of chemical reaction. The Sun's fuel is hydrogen atoms. The "burning" of hydrogen atoms to pro-

Our Sun as seen by the Solar and Heliospheric Observatory (SOHO). Image credit: SOHO (ESA & NASA)

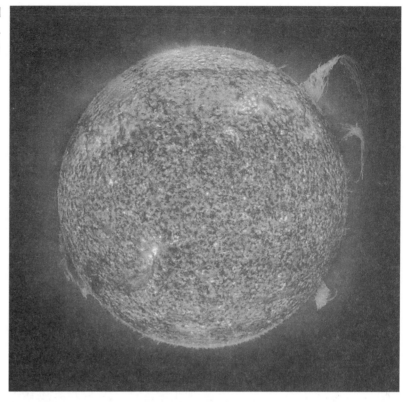

duce heat and light is done by way of nuclear reactions. Nuclear reactions are much more powerful than any chemical reaction. In the Sun's case, hydrogen atoms are fused together to form helium atoms. Millions of these nuclear reactions are going on in the Sun's core all the time.

Nuclear fusion requires very special conditions before it can occur. In fact, there is only one place in our solar system where nuclear reactions occur naturally—the center of the Sun. For these reactions to take place, there must first be a source of hydrogen. Since 75 percent of the Sun is made up of hydrogen, that requirement is the easiest to meet. A second requirement is a temperature of several million degrees. You also need extremely high pressures, pressures millions of times greater than what we are used to on Earth. Since the Sun's core has plenty of hydrogen *and* the highest temperatures and pressures in the solar system, this makes it the perfect place for nuclear reactions. This same process occurs in all stars.

When hydrogen fuses into helium, the process releases a tremendous amount of energy. However, because matter is so dense near the Sun's core, the newly released energy actually takes several thousand years to make its way to the surface and then on out into space.[5] Why? There is so much stuff in its way, it keeps bouncing into things and changing direction. It's like trying to cut across a crowded hallway. There are so many people that you bounce off one, step around another, and move from side to side to avoid someone else. Eventually you make it across the hall, but it took some extra time and it certainly wasn't in a straight line.

The energy released at the Sun's core has to travel through an incredibly crowded layer of hydrogen atoms before it can get to the surface. As a result, it gets bounced around for thousands of years. When it finally makes it to the surface, it takes less than eight and a half minutes to reach Earth.

How hot is the Sun?
It depends on which part of the Sun you are talking about.
Temperatures range from thousands to millions of degrees.

The surface of the Sun is about 10,000°F (5,500°C).[6] Sunspots—dark areas on the Sun's surface—are cooler, only 8,000°F (4,000°C).[7] As you go deeper into the Sun, the temperature rises. At the core, the temperature is several million degrees.

What is the solar wind?
A stream of charged particles constantly flowing away from the Sun.

Unlike winds on Earth (which are random movements of large quantities of atmosphere), the solar wind is made up of charged particles, mostly protons and electrons. These particles are continually pouring out of the Sun's upper atmosphere and flowing into space.

When the solar wind hits Earth, the charged particles interact with our planet's magnetic field and are channeled toward the North and South Poles. There, they collide with atoms in our atmosphere and create the northern and southern lights—the aurora borealis and aurora australis—beautiful dancing lights that illuminate the night sky.

Astronomers aren't sure how far the solar wind "blows." Its effects can be felt well past Pluto. On November 5, 2003, *Voyager 1*—the most distant human-made object—was still detecting the solar wind at a distance of 8.4 billion miles (13.5 billion kilometers) from the Sun, although the spacecraft had begun to detect a slight change in its composition.[8] That is one strong wind!

What is a sunspot?
A dark, "cool" area on the Sun's surface.

Sunspots are dark and cool only when compared to the rest of the Sun's surface. If you could magically lift a sunspot off the Sun and move it to another part of space, you would find that it is still very bright and hot. Typically, a sunspot is around 8,000°F (4,000°C).

A detailed, close-up image of a sunspot group can be found in the color insert section of this book.

Sunspots are caused by knots in the Sun's ever-changing magnetic field. On Earth, we don't have this problem because our planet's magnetic field is embedded in the solid crust of our planet. It has no choice but to rotate at the same rate as the surface of Earth, once every twenty-four hours. The Sun, however, is not solid but gaseous. Portions of it rotate at different speeds—a process called differential rotation. (Think about stirring a cup of coffee. The coffee isn't solid so portions of it will rotate at different speeds around the cup.) In the case of our star, gases at the Sun's equator take about twenty-five days to complete one rotation, while gases at the poles take almost thirty-five days to rotate once.[9] The Sun's magnetic field gets pulled along at these different rates and gets twisted in the process.

As the magnetic field becomes more and more twisted, knots begin to form. Where a knot forms, the magnetic field is very strong. It acts like a giant hand pushing down on the Sun's surface and keeps the hot gases from rising in that area. That area cools off and grows darker.[10] Thus, a sunspot is formed. Sunspots can last anywhere from a few hours to a few weeks. Eventually, however, the knot will relax and the sunspot will dissipate.

Are there always sunspots on the Sun?
No.

The Sun goes through an eleven-year sunspot cycle.[11] At the beginning of the cycle, called solar minimum, the Sun's magnetic field is aligned with its north and south poles, so there is no twisting or knotting of the magnetic field and therefore no sunspots. However, this alignment doesn't last for long.

The Sun's rotation slowly begins to twist the magnetic field. At first, only a few sunspots appear, but the numbers begin to increase as the magnetic field becomes more and more twisted. Sunspot activity reaches its peak five and a half years into the cycle. This peak is called solar maximum. Sunspots can be found on the Sun almost every day during this time. After the peak, the magnetic field, in effect, begins to unwind. For the next five and a half years, the number of sunspots will decline until there are none, and the cycle begins again.

There is a second sunspot cycle that is twice as long as the one explained above. During solar maximum, the polarity of the Sun's magnetic field flips. North becomes south and south becomes north. Then, eleven years later, at the next solar maximum, the process repeats itself with the magnetic poles flipping again. So it takes two sunspot cycles, or twenty-two years, to return to the same magnetic polar alignment.[12]

What is a solar flare?
A sudden release of energy from the Sun's surface caused by the snapping (or breaking) of the Sun's magnetic field.

Occasionally, stresses on the Sun's knotted magnetic field prove to be too great and the field simply snaps, producing a solar flare. During a flare, large amounts of charged particles from the solar surface are flung into space over a very short period of time. Flares are directly related to sunspot activity. Where there are sunspots, there are magnetic fields under stress. As a result, the number of solar flares increases during solar maximum and decreases during solar minimum.

A less well-known and less powerful type of solar explosion is an erupting prominence. Prominences are cloudlike features that form just above the Sun's surface, usually near sunspots (where the Sun's knotted magnetic field extends above the surface). Gases within the atmosphere are compressed by the magnetic field until they heat up and glow. Occasionally, a magnetic field snaps and the heated atmospheric gases are released. These eruptions are not nearly as powerful as solar flares, but they provide some great photo opportunities.

A detailed image of an erupting prominence can be found in the color insert section of this book.

If Earth happens to be in line with the stream of charged particles released by a solar flare, the stream can bombard our planet, sometimes causing disruptions in communications. While the effects of a solar flare on Earth can be bothersome, they pale in comparison to the effects of a coronal mass ejection.

What is a coronal mass ejection?
A tremendous release of energy from the Sun's upper atmosphere (its corona).

A coronal mass ejection is much more powerful than a solar flare and as a result can have a much more dramatic effect on our planet. Unlike flares, these enormous ejections of mass originate high in the Sun's corona (upper atmosphere) and can occur at any time, even during the relatively quiet time on the Sun's surface known as solar minimum. The forces behind these ejections are not completely understood, but it is believed that they are related to the Sun's ever-changing magnetic field.

Whatever the cause, something within the Sun's upper atmosphere triggers the violent release of huge bubbles of magnetically charged gas from the corona. These bubbles contain anywhere from one to ten billion tons of solar wind. During solar maximum, coronal mass ejections occur between two and three times a day. During solar minimum, the rate slows to about one a week.[13] Even though they occur fairly frequently, they don't always affect Earth because they are not always aimed in our direction. If the matter ejected during one of these explosions does happen to be aimed in the direction of our planet, things can get interesting. And that is exactly what happened in late October 2003.

An extremely large coronal mass ejection erupted from the Sun's upper atmosphere on October 22, 2003—even though the Sun was well on its way to its 2005 solar minimum period.[14] Particles released from the massive solar explosion reached Earth on October 24, 2003, and caused satellites to malfunction, triggered cell phone communications to break down, required airplanes in the North Atlantic to alter course, and even forced astronauts onboard the International Space Station into radiation-protected areas of the station while in view of the Sun.[15]

The coronal mass ejection even interfered with spacecraft on their way to Mars. The intense flood of charged particles interacted with the star cameras on both *Spirit* and *Opportunity*, NASA's two Mars Exploration Rovers. Some of the charged particles registered as false stars on each spacecraft's star camera. These extra stars didn't match any of the star maps used for navigation, and as a result, both spacecraft became slightly confused. Engineers were able to clear the star cameras, but as a precaution, NASA chose to reboot the entire systems, just in case the storm had affected other components of the spacecrafts' systems that they were unable to detect. The reboot was successful, and each rover went on to land successfully on the Red Planet.[16]

The storm's effects were even felt (or in a sense, heard) by the *Cassini* spacecraft on its way to Saturn. *Cassini*'s instruments recorded the sounds of the charged particles as they passed by the spacecraft.[17] So, the effects of a coronal mass ejection can be felt throughout the solar system.

What is the Sun made of?
Very hot gases—specifically hydrogen and helium.

Seventy-five percent of the Sun's matter is hydrogen and 24 percent is helium. The remaining 1 percent consists of argon, calcium, chromium, iron, nickel, silicon, sulfur, and other elements.[18]

If the Sun is made of gas, could you fall right through it?
No.

There is nothing solid on the Sun for you to stand on. But that doesn't mean you would fall right through the Sun, either.

If you could magically survive the Sun's heat, the pressure you would encounter as you sank deeper into the solar gases would crush you long before you made it to the center of the Sun.

Have you ever dived to the bottom of the deep end of a swimming pool and noticed that your ears hurt? The deeper you go under the surface, the greater the pressure. Even under ten feet of water, you notice a definite increase in pressure. The water is squeezing you from all sides.

Deep-sea explorers require submarines and submersibles made out of extremely thick metal to survive the tremendous pressures found just a couple of miles beneath the ocean. Without the protective gear, they would be crushed.

Now, imagine trying to go 432,200 miles (696,000 kilometers) below the surface! The Sun is 864,400 miles (1,392,000 kilometers) in diameter, which means it is roughly 432,200 miles to the Sun's core. If humans have trouble surviving the pressure found just a few miles beneath Earth's oceans, how could anyone possibly survive 432,200 miles beneath the surface of our Sun? The pressure would be so great that you would be completely squashed shortly after you got started.

So even though the Sun is made of gas, there is no way anyone, or anything, could pass through it.

Will our Sun ever burn out?
Yes.

Don't lose any sleep over this fact, though. Astronomers figure the Sun will last another four and a half billion years.

If you know your car uses a gallon of gasoline for every twenty miles you drive, and you know your gas tank holds twenty gallons, you can predict that you will be able to go four hundred miles before you run out of gas (20 gallons × 20 miles/gallon = 400 miles).

Astronomers use the same reasoning to determine how long our Sun will continue to give off energy. They know nuclear reactions are taking place in the Sun, and they know how massive the Sun is. In other words, they know how much fuel the Sun has and how fast it is using that fuel. So it is possible to predict how long the Sun will fuse hydrogen before it "runs out of gas."

Astronomers believe that our Sun has been burning for almost four and a half billion years and has at least another four and a half billion years to go before it dies.[19]

Will the Sun just burn out and go black?
Yes and no.

Before the Sun fades to black, it will go through a series of expansions and contractions that will take it from an average yellow star (which it is now), to a red giant, to a yellow/white star, to another red giant, to a planetary nebula, and finally to a white dwarf. While intellectually interesting, beings on Earth won't have to worry about any of these phases. Why? Because when the first phase occurs several billion years from now, it will heat up our planet so that all the water in the oceans will boil away and any remaining inhabitants will be fried! Below is a summary of all the strange things yet to come for our Sun:

Phase 1: Nuclear reactions in the Sun's core cease

The Sun's weird contortions will begin in approximately four and a half billion years, when the Sun literally runs out of gas. The nuclear reactions in the Sun's core will eventually use up all the hydrogen. With no more fuel, nuclear fusion will stop, and the pressure and energy pushing outward from the core will cease as well. Without anything to oppose it, the gravity pushing inward from the mass of the surrounding material will take over and the core (now composed of helium) will collapse in on itself.[20]

As the core collapses, its pressure and temperature will increase dramatically, until it is even hotter than it was when nuclear reactions were taking place.[21] These conditions will trigger nuclear reactions in a hydrogen layer immediately surrounding the core and mark the beginning of the Sun's second phase of death.

Phase 2: Sun swells to form a red giant

With a hydrogen-burning layer just above its extremely hot collapsing core, our Sun will be emitting much more energy than it ever has before. As a result, it will begin to swell in size. As the Sun expands, its surface will begin to cool. From the average-sized, yellow-hot Sun we are familiar with, it will grow and cool, becoming a giant reddish-colored star. During this process, the Sun will overtake and swallow the planets Mercury and Venus. When the expansion finally stops, the Sun's surface will almost reach the orbit of Earth.[22] Even though its surface will be almost 3,600°F (2,000°C) cooler than it was before, this drop in temperature will not affect Earth. In fact, because the Sun's surface will now be much closer to our planet, Earth will be scorched. All the water in the oceans will boil away and conditions for any remaining life-forms will be pretty grim, to say the least.

At this point, the Sun could be officially classified as a red giant star.[23] It will remain a red giant for about three billion years.[24] All during this time, the hydrogen-burning layer surrounding the core will be producing helium. This helium will fall onto the collapsing core, increasing its mass. As more mass is added to the core, the greater its pressure and temperature will become. Finally, after three billion years, the pressure and temperature in the core will be great enough to force helium atoms to fuse together to form carbon, thus marking the beginning of helium fusion in the core and the Sun's third phase of death.

Phase 3: Helium fusion begins in the Sun's core

The collapse of the Sun's core will immediately stop when helium fusion begins. Without the tremendous heat generated from the core's collapse, the temperature of the surrounding gases will cool slightly, bringing a halt to the hydrogen burning in the layer above the core.[25] Now the sole source of the Sun's energy will be the helium fusion burning in its core. The core will expand as energy from the helium fusion is released. To counter the core's expansion, the surface of the Sun will contract, returning it almost to its original size.

The helium-burning phase of the Sun's life will not last nearly as long as the hydrogen-burning phase. For one thing, there won't be nearly as much helium. (The process of hydrogen fusion requires four hydrogen atoms to make one helium atom.) Helium reactions also occur at a much more rapid pace because of the higher pressure and temperature required for the reactions to take place. As a result, the Sun will be burning at a slightly hotter temperature than it was during our lifetime. The helium-burning phase will last about one hundred million years.[26] (This may seem like a long time based on human standards, but for a star's life measured in billions of years, it really isn't much time.)

Once the nuclear reactions have fused all of the helium into carbon and oxygen, the reactions will stop and the core will begin to collapse again, just as it did after the hydrogen-burning stage was complete. As the Sun starts to swell, the fourth phase of the Sun's death will begin.

Phase 4: Surface of Sun swells to form a red giant

With temperatures and pressures increasing because of the collapsing core, more nuclear reactions will be triggered in layers surrounding the core. This time a layer of helium—left over from the first hydrogen-burning layer—will begin fusing into carbon and oxygen. At the same time, a layer of hydrogen above it will begin to fuse.[27] So, not only will there be a tremendous amount of energy being released by the collapsing core, there also will be energy from two fresh layers of nuclear reactions. With energy pouring out of the star as never before, the Sun will expand once again to form a red giant. This time, however, because of the tremendous amount of energy being released, the Sun will be a red giant for only a few million years. The end of the red giant phase will mark the beginning of the fifth phase of the Sun's death.

Phase 5: Our Sun becomes a planetary nebula
(an old star surrounded by an expanding shell of gas and dust)

The second red giant phase will be an extremely unstable time for the Sun. The helium layer just above the collapsing core will burn unevenly, releasing pulses of energy. These pulses will eject large amounts of the Sun's atmosphere into space.[28] The former atmosphere will move away from our dying star as an expanding shell of dust and gas. At this point, our Sun will become a planetary nebula. The planetary nebula phase will continue for approximately fifty thousand years.[29]

When planetary nebulae were first observed in the early 1800s, astronomers weren't sure what they were seeing. The fuzzy disks of these mystery objects resembled the planets of Uranus and Neptune, but their positions remained fixed in the sky, like that of a distant star. For lack of a better name, astronomers called this new class of objects planetary nebulae. Today, even though we know that a planetary nebula has nothing to do with planets, the name remains.

A colorful example of a planetary nebula can be found in the color insert section of this book.

After fifty thousand years, as the shell of gas and dust that classifies our Sun as a planetary nebula dissipates into interstellar space, all that will remain of our star is the hot collapsing core. This will mark the sixth and final phase of the Sun's death.

Phase 6: The Sun collapses into a white dwarf

As the core of the Sun continues to collapse in on itself, its temperature and pressure will continue to rise. This time, however, the Sun won't have enough mass for the next round of nuclear reactions to begin. With no more fusion reactions occurring in the core, the star will be considered dead. When the collapse eventually slows to a stop, what is left of our Sun will be a sphere about the size of our planet.[30] At this point, the Sun will be officially a white dwarf—an extremely small, exceedingly dense, dead star. Because of the tremendous heat it generated during its collapse, our white dwarf Sun will continue to glow for billions of years before it eventually fades to black.

So this brings to an end the long and complicated answer to the simple question "Will the Sun just burn out and go black?" Are you sorry you asked?

What is a solar eclipse?
What happens when the Moon passes directly between Earth and the Sun.

Occasionally during its New Moon phase, the Moon's orbit takes it directly between Earth and the Sun. As this happens, the Moon slowly covers and then uncovers part or all of the Sun's disk.

There are three types of solar eclipses: a partial eclipse, a total eclipse, and an annular eclipse.

The Sun just beginning to peak out from behind the Moon after a total eclipse—as seen from Aruba on February 26, 1998. Image credit: Dr. Philip Knocke

A partial solar eclipse occurs when the Moon passes slightly above or below the Sun, so it covers only a portion of the Sun's disk.

A total solar eclipse occurs when the Moon passes directly between the Sun and Earth and completely covers the Sun's disk for a short period of time.

An annular solar eclipse occurs when the New Moon is at its farthest point from Earth. Because the Moon is farther from Earth than usual, its disk appears slightly smaller in our sky. As a result, when the Moon passes in front of the Sun, it isn't large enough to completely cover the Sun's disk. Instead, it creates a bright donut of sunlight when the dark Moon passes across the Sun's center. If the Moon were closer to Earth during an annular eclipse, it would be a regular total eclipse.

Why doesn't a solar eclipse happen every month—with each New Moon?
Because the Moon's orbit is tilted with respect to Earth and the Sun. Most of the time, the New Moon is either above or below the Sun's position in our sky.

If you were to fly above our solar system and look down, you would see our Moon circling Earth once every twenty-nine and a half days. From this vantage point, as shown in the following illustration, it would appear that every time the Moon moves between Earth and the Sun, there should be an eclipse.

Overhead View (not to scale)

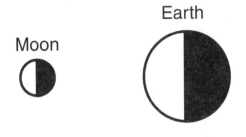

However, if you were to view our solar system from the side, as shown by the illustration below, you would see that the Moon's orbit is tilted with respect to Earth and the Sun. For much of the time, this tilted orbit takes the Moon well above or below the Sun's location in our sky during its New Moon phase. Only on occasion does the New Moon actually pass in front of the Sun, causing an eclipse.

Side View (not to scale)

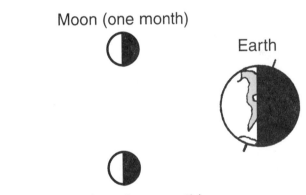

When is the next solar eclipse?

Listed in the chart below are the dates, times, and best viewing locations for all the partial and total solar eclipses through the year 2010.
For a list of solar eclipses through the year 2020, see page 206.

WARNING!—Viewing the Sun can be extremely dangerous!

If you want to look at the Sun, use only specially designed filters.
Observing the Sun without proper protection can cause permanent blindness!

Solar Eclipses for 2005–2010

Date	Eclipse Type	Time of Maximum Coverage*	Duration of Total Phase	Partial eclipse visible from:	Total or Annular eclipse visible from:
2005 Apr 08	Annular/ Total	20:36 UT (4:36 PM EDT)	00m42s	New Zealand, Americas	southern Pacific, Panama, Colombia, Venezuela
2005 Oct 03	Annular	10:32 UT	04m32s	Europe, Africa, southern Asia	Portugal, Spain, Libya, Sudan, Kenya
2006 Mar 29	Total	10:11 UT	04m07s	Africa, Europe, western Asia	central Africa, Turkey, Russia
2006 Sep 22	Annular	11:40 UT	07m09s	South America, western Africa, Antarctica	Guyana, Suriname, French Guiana, southern Atlantic
2007 Mar 19	Partial	02:32 UT	—	Asia, Alaska	—
2007 Sep 11	Partial	12:31 UT	—	South America, Antarctica	—
2008 Feb 07	Annular	03:55 UT	02m12s	Antarctica, eastern Australia, New Zealand	Antarctica
2008 Aug 01	Total	10:21 UT (6:21 AM EDT)	02m27s	northeastern North America, Europe, Asia	northern Canada, Greenland, Siberia, Mongolia, China
2009 Jan 26	Annular	07:58 UT	07m54s	southern Africa, Antarctica, Southeast Asia, Australia	southern Indian Ocean, Sumatra, Borneo
2009 Jul 22	Total	02:35 UT	06m39s	eastern Asia, Pacific Ocean, Hawaii	India, Nepal, China, central Pacific
2010 Jan 15	Annular	07:06 UT	11m08s	Africa, Asia	central Africa, India, Myanmar, China
2010 Jul 11	Total	19:33 UT	05m20s	southern South America	southern Pacific, Easter Island, Chile, Argentina

*Times given in Universal Time (UT) and Eastern Standard Time (EST) or Eastern Daylight Time (EDT) where appropriate.

The author would like to thank Fred Espenak, NASA Goddard Space Flight Center, for the use of his eclipse charts.[31]

Why shouldn't you look directly at the Sun?
Because the Sun's radiation can damage your eyes before you are aware of it.

Yes, the Sun's brightness can blind you. However, it's not just the intensity of light that can damage your eyes. It's the stuff that you can't see that can seriously harm you. If you were to use a pair of sunglasses or other dark glasses to look directly at the Sun, it might seem safe, but invisible radiation could be passing through the glasses and burning your eyes without your knowledge. Since there are no pain sensors in the backs of your eyes, you wouldn't know that any damage was being done until it was too late.

The invisible, damaging energy from the Sun is ultraviolet radiation. Undetected by the human eye, ultraviolet light can pass through clouds and most so-called filters. (Ultraviolet radiation is what causes sunburns on a cloudy day.) Don't take any risks by looking at the Sun through a filter unless you have utter confidence in the filter's source and quality.

Many museums and public observatories offer safe solar viewing. Nature also provides a safe way to view a solar eclipse without any special devices. Just go outside at the time of the eclipse and look at the ground (not the sky!) beneath a shady tree. In the right kind of shady area, you will be able to see *on the ground* several round images of the Sun, each with a bite taken out of it. The leaves of a tree filter the sunlight for ground viewing—forming a natural pinhole projector that projects the Sun's image onto the ground. Normally, we don't pay any attention to the overlapping images of the Sun beneath our feet (that dappled look of part sun, part shade). During an eclipse, however, when the Moon starts taking a bite out of the Sun, suddenly we notice the ground covered with partial circles.

The Sun is a fascinating object whose surface is in a constant state of change, but when in doubt, don't look. Your eyesight is not worth the risk.

ABOUT THE PLANETS

Where did the planets get their names?

What's an easy way to remember the names of the planets?

How far away are the planets?

How big are the planets?

Which is the biggest planet?

Which is the smallest planet?

What is a gas giant planet?

Which is the coldest planet?

Which is the hottest planet?

Which planet has the largest canyon?

Which planet has the largest volcano?

Does every planet have a moon?

Which planet has the most moons?

Which planet has the largest moon?

What would it be like to live on Mercury?

Why is Venus sometimes called the Morning Star or the Evening Star?

What would it be like to live on Venus?

Why is Mars red?

Are there canals on Mars?

Why should we care if there is water on Mars?

How many spacecraft have actually landed on Mars?

What would it be like to live on Mars?

What is the Great Red Spot on Jupiter?

Could you land a spacecraft on Jupiter?

What would it be like to live on Jupiter?

Are there oceans on some of Jupiter's moons?

What are the rings of Saturn made of?

Is Saturn the only planet that has rings?

Is Pluto really a planet?

Is there life on any other planet in the solar system?

Are there planets around other stars?

Where did the planets get their names?
From the ancient Greeks and Romans.

The word "planet" is derived from the Greek word meaning "wanderer." The ancients noticed that most of the stars in the sky stayed fixed in their constellations. However, five of these "stars" moved, or wandered across the sky. Because the ancient sky watchers didn't understand why these special "stars" moved when all others remained fixed, they decided these wanderers must have something to do with the gods. Just to play it safe, they named the planets in honor of the gods and goddesses of the time. The ancient Greeks named the planets. The Romans later adopted the Greek mythology but changed all of the characters' names to Roman names (Zeus = Jupiter, Aphrodite = Venus, etc.). Today we know the planets by their Roman names.[1]

- Mercury was named after the fleet-footed messenger god because of its rapid movement across the sky. (Being the closest planet to the Sun, Mercury takes only eighty-eight days to complete one orbit.)
- Venus was named after the goddess of beauty because its brightness dominates the early morning or early evening sky, depending on where it is along its orbit.
- Mars was named after the god of war because of its blood-red appearance.
- Jupiter was named after the king of the gods. (Keep in mind the ancient Greeks and Romans had no way of knowing that Jupiter is by far the largest planet in the solar system.) Jupiter's brightness does not equal that of Venus, but it seems to dominate the heavens as its orbit takes it slowly across the night sky.
- Saturn, the slowest moving of the visible planets, was named after the king of the Titans (a mythical race of giants), who was also the god of agriculture. Saturn was the son of Uranus and the father of Jupiter.

Mercury, Venus, Mars, Jupiter, and Saturn were the only planets named by the ancients because they were the only ones they could see. The more distant planets—Uranus, Neptune, and Pluto—are visible only with a telescope and were therefore not discovered until much later. (Uranus was discovered in 1781, Neptune in 1846, and Pluto in 1930.) When naming the more distant planets, astronomers tried to continue in the ancient tradition. Uranus was a son of Gaia (Mother Earth) and one of the first gods. He was also the father of Saturn and the grandfather of Jupiter. Unlike all of the other planets, Uranus was named after a Greek god. He has no Roman counterpart. Neptune was god of the seas and Pluto was the ruler of the underworld.

What's an easy way to remember the names of the planets?
That depends on your memory.
Getting to know the planets as individual worlds is the best way.
However, you may try using one of the memory aids listed below.

Knowing the planets is one thing. Trying to name the planets with only a moment's notice is something else entirely. Unless you know something about each planet that makes it its own unique world, trying to remember the names and order of the planets is like trying to remember the names of the seven dwarfs.

The nine planets in our solar system are Mercury, Venus, Earth, Mars, Jupiter, Saturn, Uranus, Neptune, and Pluto—listed in order, beginning with the closest planet to the Sun. Below are a couple of phrases that might help you remember the order of the planets. Each phrase uses the first letter of each planet to create its own crazy sentence. The idea being that it is easier to remember one wacky sentence than nine individual names.

Mercury,	Venus,	Earth,	Mars,	Jupiter,	Saturn,	Uranus,	Neptune,	Pluto
My	Very	Educated	Mother	Just	Served	Us	Nine	Pizzas.
Many	Vampires	Enjoy	Mango	Juice	Sitting	Under	Nordic	Pines.

Use one of these two phrases, or make up your own. The planets won't change their orbits, so once you memorize this order, you are set. (There is a twenty-year period when Pluto is closer to the Sun than Neptune is, but that won't happen again until the year 2227.)

How far away are the planets?

Planet	Average Distance from the Sun (in miles)	Average Distance from the Sun (in kilometers)[2]
Mercury	35,950,000	57,900,000
Venus	67,190,000	108,200,000
Earth	92,900,000	149,600,000
Mars	141,500,000	227,900,000
Jupiter	483,300,000	778,300,000
Saturn	887,400,000	1,429,000,000
Uranus	1,785,000,000	2,875,000,000
Neptune	2,982,000,000	4,504,000,000
Pluto	3,674,000,000	5,916,000,000

Because all of the planets orbit around the Sun at different rates, their distances from each other are constantly changing. As a result, planetary distances are usually given with respect to the Sun. These distances are so vast that they are sometimes hard to comprehend. So instead of miles or kilometers, let's describe the solar system in terms of travel time.

It took Apollo astronauts three days to get to the Moon—and that is the closest object to our planet.[3]

It took *Mariner 10* three months to reach Venus and almost five months to reach Mercury.[4]

It took the Mars Exploration Rovers *Spirit* and *Opportunity* seven months to get to Mars.[5]

It took *Voyager 2* a year and a half to get to Jupiter, three years to get to Saturn, eight and a half years to get to Uranus, and twelve years to get to Neptune.[6]

No spacecraft has traveled to Pluto.

How big are the planets?

Planet	Equatorial Diameter (in miles)	Equatorial Diameter (in kilometers)[7]
Mercury	3,030	4,880
Venus	7,515	12,102
Earth	7,921	12,756
Mars	4,219	6,794
Jupiter	88,793	142,984
Saturn	74,853	120,536
Uranus	31,744	51,118
Neptune	30,757	49,528
Pluto	1,430	2,300

Which is the biggest planet?
Jupiter.

Jupiter is by far the largest planet in the solar system. Seventy percent of all the mass in the solar system—outside of the Sun, that is—is found within Jupiter.

Jupiter is over eleven times larger than Earth, which means you could fit eleven Earths across Jupiter's diameter. If Jupiter were hollow, like a giant gumball machine, it would hold fourteen hundred Earth-sized gumballs.

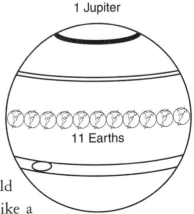

1 Jupiter

11 Earths

The size of Earth compared with Jupiter.

Which is the smallest planet?
Pluto. (Yes, it's still officially a planet.)

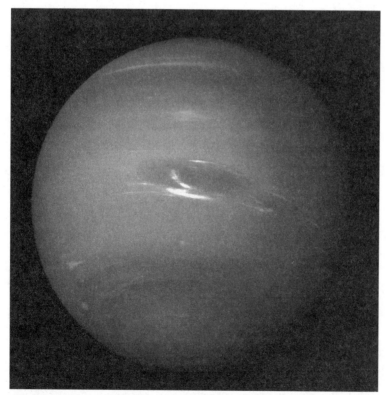

The most distant planet is also the smallest. Pluto is even smaller than Earth's moon. This tiny planet is 1,430 miles (2,300 kilometers) in diameter while our Moon is 2,158 miles (3,476 kilometers) across.[8]

Because of its distance from the Sun—almost 4 billion miles (6.6 billion kilometers)—and its small size, Pluto was not discovered until 1930. (It was discovered by Clyde Tombaugh, a young astronomer working at Lowell Observatory.)[9] The ninth planet cannot be seen with the naked eye. Even through a telescope, Pluto appears only as a faint spot. (For more about Pluto, see "Is Pluto really a planet" on page 67.)

The size of our Moon compared with Pluto.

Neptune is just one of four gas giant planets in our solar system.
Image credit: NASA/JPL

What is a gas giant planet?
A large planet made up mostly of hydrogen and helium gas.

There are four gas giant planets in orbit around our Sun: Jupiter, Saturn, Uranus, and Neptune. Gas giant planets are the largest planets in the solar system and are found farther from the Sun than most planets. They also have more moons and are circled by complex ring systems of varying size and composition.

Since gas giants are composed almost entirely of gas, they have no solid surface on which to land. A spacecraft trying to "land" on a gas giant would find itself passing through denser and denser clouds until it is eventually crushed by the pressure of the thickening gas. Each gas giant planet does have a small solid core at its center, but this core cannot be considered a rocky surface. For example, Jupiter's core is believed to be about the size of Earth. While that may sound large, remember that over fourteen hundred Earths could fit inside Jupiter. To get to that small rocky core, one would have to travel down through 40,436 miles (65,114 kilometers) of Jupiter's gases, encountering incredible pressures along the way. There is no way that any human-built spacecraft could survive such a journey—at least not with current technologies. The *Galileo* probe, which plunged into Jupiter's upper cloud layer on December 7, 1995, only made it to a depth of 120 miles (200 kilometers) before it was crushed by pressures twenty-four times greater than Earth's pressure at sea level and melted by a temperature of 305°F (152°C)—and that was only one hour into its plunge![10]

Which is the coldest planet?
Pluto.

It makes sense that the most distant planet from the Sun is also the coldest. With an average temperature of –387°F (–233°C), Pluto is frigid![11]

Which is the hottest planet?
Venus.

With a planetwide average temperature of 900°F (480°C), Venus holds the record for being the hottest planet in the solar system.[12] This is strange, if you think about it. After all, Mercury is closer to the Sun than Venus. Why isn't Mercury the hottest planet? It has to do with atmosphere: Mercury has none and Venus has too much.

On Mercury, the lack of an atmosphere means that heat trapped by surface rocks during the day quickly radiates into space at night. There is nothing to block its escape. Temperatures that can get as high as 660°F (350°C) during the day plummet to –270°F (–170°C) at night.[13]

On Venus, its extremely dense atmosphere allows sunlight through to warm the planet's surface. But when the resulting heat rises and tries to escape back into space, its path is blocked by carbon dioxide in the atmosphere. The gaseous carbon dioxide acts almost like a solid wall, refusing to allow the infrared radiation (heat) to pass through. Since over 96 percent of Venus's atmosphere is carbon dioxide, this forms a pretty thick wall! The infrared radiation is forced to remain on the planet, increasing its temperature. In the meantime, the surface continues to be heated by sunlight. This heat is also blocked by the atmosphere, and the surrounding temperature rises again as the process repeats itself. This is what scientists call a runaway greenhouse effect. On Venus, this effect has created a planet with an average temperature of 900°F (480°C), be it day or night, at the poles or at the equator. To help put things in perspective, you need only 450°F (232°C) to bake a pizza and 621°F (327°C) to melt lead.[14]

Which planet has the largest canyon?
Mars.

Valles Marineris, or Mariner Valley, is a huge canyon that stretches for more than 3,000 miles (4,700 kilometers) across the Martian surface. The canyon plunges to a depth of 5 miles (8 kilometers) at its deepest point and spans 400 miles (650 kilometers) at its widest.[15] For comparison, Earth's Grand Canyon covers about 250 miles (403 kilometers) of the northwestern corner of the state of Arizona and runs about a mile wide and a mile deep.[16]

Valles Marineris, the largest canyon in the solar system, cuts a wide swath across the Martian surface. Volcanoes of the Tharsis Ridge can be seen to the left.
Image credit: NASA/USGS

Mariner Valley is actually a rift valley. Unlike the Grand Canyon, which was formed as water from the Colorado River ate away at the surrounding rocks, Mariner Valley was created when the crust of its planet pulled apart. Earth has its own rift valleys, one of which we know as Africa's Lake Tanganyika, the longest freshwater lake in the world with a depth of almost a mile (1.4 kilometers).[17]

Which planet has the largest volcano?
Mars.

The largest volcano in the solar system, Olympus Mons, is found on a planet about half the size of Earth. This volcano rises 15 miles (25 kilometers) above the surrounding Martian landscape, supported by a base that is almost 400 miles (600 kilometers) in diameter.[18] For perspective, Olympus Mons is three times taller than Mount Everest, the highest point on our entire planet. Its base could completely cover the entire state of Missouri.

Olympus Mons is similar to shield volcanoes found in the Hawaiian Islands.[19] These volcanoes form over an area where molten lava from Earth's interior is forcing its way up through the planet's outer crust. Lava will continue to flow from these "hot spots," building layer upon layer, until either the area beneath the crust cools off or the crust containing the volcano is carried away from the hot spot. The latter is how the Hawaiian Islands came to be. After one island/volcano was created, plate tectonics—the movement of large portions of Earth's crust—carried it away from the hot spot. That volcano became extinct, while another began to form over the stationary hot spot. This process continues today.

On Mars, Olympus Mons began its life over a hot spot in the Martian mantle. Unlike Earth, however, Mars did not have any active plate tectonics. (There may have been plate tectonic activity early in the planet's history, but that ceased long ago.) With no force to move it away from the hot spot, this volcano continued to erupt and grow in size. It wasn't until the hot spot in the planet's interior cooled that Olympus Mons finally stopped erupting, leaving it at its record-breaking height. In addition to Olympus Mons, there are other large volcanoes on the Martian surface, several of which are found on the nearby Tharsis Ridge.

Does every planet have a moon?
No.

The two innermost planets, Mercury and Venus, have no moons. All the other planets have at least one moon in orbit around them. (Moons are often referred to as satellites.) As a general rule, the gas giant planets have multiple moons while the smaller, rocky planets have few or none.

Earth and Mars—smaller, rocky planets—have one moon and two moons, respectively. Of the gas giant planets, Jupiter has the most moons with at least sixty-three. Saturn is next, with thirty-three moons. Uranus has twenty-seven moons, Neptune has thirteen, and Pluto has one.[20]

These totals represent the number of known moons—as of the printing of this book. However, they probably don't represent the total number of moons out there. In the last ten years, seventy-eight moons have been added to the list, most of them in the last four years. With better observing techniques and more sensitive equipment, no one knows how many more moons astronomers will find.

Which planet has the most moons?
Jupiter.

Jupiter has sixty-three known moons, and there may be more that astronomers haven't found yet.

A stunning image of Jupiter's moon Io, with its enormous planet dominating the background, can be seen in the color insert section of this book.

It makes sense that Jupiter, the largest and most massive planet in the solar system, has the most moons. Such a massive planet has an incredibly strong gravitational field. This strong gravity attracts stray asteroids and comets that pass too close to the planet. By altering their orbits, Jupiter's gravity attracts these strays and brings them into the Jovian family. Most of Jupiter's sixty-three moons are considered "captured" moons, moons not originally formed with the gas giant but snared and kept in orbit by the planet's gravity.

Jupiter's four largest moons—Io, Europa, Ganymede, and Callisto—are believed to have been formed with the planet. These moons were the first astronomical objects to be discovered using a telescope—they were discovered by Galileo in 1610![21] Other Jovian moons include Metis, Adrastea, Amalthea, Thebe, Leda, Himalia, Lysithea, Elara, Ananke, Carme, Pasiphae, and Sinope.[22] (For more moon names, see page 203.)

Which planet has the largest moon?
Jupiter.

Ganymede, the largest moon of Jupiter, is also the largest moon in the solar system. It is 3,268 miles (5,262 kilometers) in diameter.[23] Thus Ganymede is larger than the planets Mercury at 3,030 miles (4,880 kilometers) and Pluto at 1,430 miles (2,300 kilometers). For comparison, our Moon is 2,158 miles (3,476 kilometers) in diameter. (For a list of other moons in the solar system ranked by size, see page 205.)

Ganymede, in orbit around Jupiter, is the largest moon in the solar system. Image credit: NASA/JPL

The surface of Ganymede can be described very simply as light and dark. About one-third of the moon is covered by dark, fractured, cratered terrain, while the rest of the surface is covered by bright, relatively young terrain.

Craters can be found on both the light and the dark terrain—although there are more on the older, darker surface. Some larger craters appear as white marks on the moon. The white coloring suggests that when meteors slammed into Ganymede, they broke through a thin dusty layer covering the moon and revealed a white water-ice layer below.

The interior of Ganymede is made up of roughly equal amounts of rock and ice. Beneath a 90-mile (150-kilometer) layer of ice, there is evidence to suggest the existence of a saltwater ocean.[24] While liquid water is one criterion for life as we know it, Ganymede's ocean—if it does exist—is probably too cold and dark to harbor any life-forms. Two other moons of Jupiter, Callisto and Europa, also appear to have oceans beneath their icy crusts.[25] (For more information, see "Are there oceans on some of Jupiter's moons?" on page 65.)

What would it be like to live on Mercury?
*It would be very hot during the day, very cold at night,
and there wouldn't be much to look at but rocks, craters, and more rocks.
Oh, and you would have to bring along your own air since Mercury has none.*

If you stood on the surface of Mercury, everything you would see and feel would be the direct result of one thing—the planet's lack of atmosphere. The surface has craters everywhere you'd look since without an atmosphere to protect it, every piece of space debris aimed at Mercury slams into the planet and forms a crater. (On other planets, friction between an atmosphere and a meteor causes most meteors to burn up before they can make it down to the surface.) Mercury's daytime sky would be black, except for the bright ball of the Sun. There would be no atmosphere to diffuse the sunlight.

Without an atmosphere to block dangerous solar radiation, Mercury is bombarded by all types of nasty things. Standing on the surface, you would be baked by infrared radiation that heats the surface to temperatures as high as 660°F (350°C). You would also be flooded by large amounts of ultraviolet radiation, x-rays, and gamma rays, leaving you and the entire surface of the planet sterilized—and dead, since no life-forms we know of could survive that much radiation!

Mercury, a barren, cratered-covered world, is the closest planet to the Sun. Image credit: NASA/JPL

Then, as the planet rotates from day to night, all the heat that was baked into the rocks during the day quickly radiates into space. With no atmosphere to trap the heat, temperatures plummet to –270°F (–170°C). So between midday and midnight on Mercury, you would have to withstand a difference in temperature of 930°F (5000°C). Ouch!

If you could magically survive the lack of atmosphere, the harsh radiation, and the dramatic changes in temperature, you would still have to endure some really long days and nights. Normally, the definition of a day is the amount of time it takes a planet to rotate once around on its axis. On Mercury, it's not that simple. Because Mercury's rotational period is equal to two-thirds of its orbital period, it actually takes two orbits (two years) and three rotations (three days) for the planet to go from sunrise to sunrise.[26] In other words, one full day and night on Mercury is equal to 176 Earth days. That would be quite a day to plan!

While conditions on its surface would be difficult to deal with (to say the least!), it may still be possible for humans to visit Mercury someday—given enough protective shielding and insulation, of course. For now, however, we will continue to study Mercury from afar.

Why is Venus sometimes called the Morning Star or the Evening Star?
Because when visible, Venus is the brightest starlike object in the night sky.

Venus's brightness in our skies can be attributed to three things: the fact that it is the closest planet to our own, the fact that the planet is covered by a thick layer of highly reflective clouds, and the fact that it is the second planet from the Sun and therefore receives a large amount of sunlight to reflect. The result is an extremely bright starlike object in our early evening sky or early morning sky, depending on where Venus is along its orbit. Only the Moon and Sun are brighter.

The reason we only see this bright planet in the early evening or early morning is because Venus is closer to the Sun than Earth is. From our vantage point on Earth, Venus is never far from the Sun. So we either see it just after sunset following the sinking Sun to the west, or we see it rising in the east just before sunrise.

Thick clouds hide the surface of Venus. Image credit: NASA

What would it be like to live on Venus?
Very hot, very dry, and very crushing.

Venus would be a really nasty place to live (for us Earthlings anyway). With temperatures hot enough to melt lead and pressures strong enough to crush all but the most heavily reinforced spacecraft, humans visiting this world would find conditions on the planet second from the Sun next to impossible to survive.

When standing on Venus, it would be a toss-up as to what you'd notice first, the scorching heat or the crushing pressure. With a planetwide average temperature of 900°F (480°C), there is nowhere to go to escape from the heat. Fortunately, it's a dry heat—a very dry heat! Since it takes a temperature of only 212°F (100°C) to boil water, all of the water on Venus boiled away a long time ago.

So maybe you anticipated the intense heat and brought a cooler full of your favorite drinks. Unless the cooler is constructed of super high-strength, heat-resistant materials, the scorching temperature would melt it as the crushing pressure collapsed it into a tiny, unrecognizable lump. The cooler wouldn't be the only thing affected by the pressure. You would also feel its crushing effects in a less-than-pleasant way.

Atmospheric pressure is something we don't think much about on Earth. Since we were born here, we don't really notice that the air we are breathing is also pushing in around us with a force of 14.7 pounds per square inch.[27] (Something to think about the next time you lift a fourteen- or fifteen-pound bowling ball.) On Venus, the atmospheric pressure is ninety times greater than on Earth.[28] You would definitely notice the difference. The Venusian air would press around you with a force of 1,323 pounds per square inch. Talk about getting squeezed!

Now, let's say you could magically survive the heat and pressure. The view you would see on the surface of Venus would be rather dark and overcast. Since Venus is perpetually covered in clouds, you would never be able to see the Sun during the day or any stars at night. While solid and constant, these clouds never produce any moisture that makes it to the surface. The intense heat of the planet causes any moisture to evaporate long before it hits the ground. As strange as it may sound, this may actually be a good thing. The clouds of Venus are not made of water vapor like clouds on Earth. Instead, they are composed almost entirely of concentrated sulfuric acid and other nasty corrosive compounds that can eat through a variety of metals as well as human flesh. If it did rain on Venus, it would not be a nice refreshing shower. It would be more like a bath of hot corrosive acid.

As for the surrounding desertlike terrain, low lying hills and lava domes dominate the landscape in some areas. In other places you might find a series of cracks or grooves running across a flat surface. There are also several volcanoes on the planet, but scientists can't tell if they are active, dormant, or even extinct. Radar images provide scientists with the detailed shapes of objects on the surface, but they can't delineate whether a volcano is slowly belching smoke and rocks or not.

Even if volcanoes are not actively belching hot noxious gases into the already nasty atmosphere, conditions on the surface of Venus are so bad that the human race should probably look elsewhere when searching for a new world to colonize.

Why is Mars red?
It's rusting.

The surface of Mars contains a small amount of iron. The Martian atmosphere contains a small amount of oxygen. The iron on the surface oxidizes as it interacts with the oxygen. Or, to put it simply, the surface of Mars is rusting like an old car.

A picture of Spirit's *lander showing a good example of the rusty Martian landscape can be seen in the color insert section of this book.*

Are there canals on Mars?
No.

In 1877 an Italian astronomer, Giovanni Schiaparelli, reported that he saw "canali" on the Red Planet.[29] In Italian, the word "canali" means channels or grooves. Unfortunately, canali was translated incorrectly into English as canals.

When one announces the discovery of channels or grooves on a planet, this isn't likely to make headline news. However, the discovery of "canals" captured the imagination of the public. Canals brought to mind human-made trenches. This led to speculation that there were intelligent creatures living on Mars. Some went as far as suggesting that the Martian civilization was suffering from a planetwide drought and the Martians were using these canals to channel water from the melting polar ice caps to their dying cities. By the early 1900s, an American astronomer, Percival Lowell, claimed he saw as many as 437 canals crisscrossing the Martian surface.[30]

Despite Lowell's enthusiasm and detailed sketches, few astronomers shared his belief in the existence of canals. When the first spacecraft flew by the planet in the 1960s, no canals could be seen. More detailed photographs have found no sign of these mysterious canals. The canals may have been optical illusions caused by Earth's atmosphere. When observing Mars through a telescope, a viewer can see dark areas on some parts of the surface. These areas are regions with dark surface markings, not canals.

Why should we care if there is water on Mars?
If there is or was liquid water on Mars,
a variety of life-forms may have been able to exist on the planet.

Without liquid water, life as we know it cannot exist. The Mars we know today cannot support liquid water on its surface. Not only is it too cold, but the pressure of the atmosphere is too thin. Any water released on the Martian surface today would either freeze or evaporate almost instantly. However, data taken from orbiting spacecraft and surface rovers have shown features and mineral deposits that suggest Mars wasn't always so dry.

In addition to a multitude of orbital images showing what appear to be dried river beds, gullies, and flood plains crisscrossing the Martian surface, the Mars Exploration Rover *Opportunity* discovered both chemical and physical evidence that suggests its landing site on Meridiani Planum used to be home to a shallow, long-standing, salty sea.[31] Chemically speaking, what *Opportunity* discovered was high levels of sulfates in the rocks near its landing site. While high sulfate levels may not sound like much, rocks on Earth containing similar amounts of sulfates were either formed in salt water or had been exposed to salt water for a long period of time. (For example, rocks at the bottom of the ocean will slowly, over long periods of time, absorb salt from the surrounding water. The longer the rock remains in the salty water, the more salt it will collect.) While there may have been other ways for these salty rocks to form on Mars, submersion in salt water seems to be the most logical conclusion, especially when you combine it with the other evidence listed below.

Physical evidence of past water on Mars was gathered using *Opportunity*'s panoramic camera and microscopic imager. Specifically, what *Opportunity* found was vugs, spherules, and crossbedding. Vugs are a series of small, random indentations in the rocks. On Earth, vugs are found in rocks that were submerged in salty water for long periods of time. Salt crystals trapped in these rocks slowly dissolve, leaving behind a hollow indention (a vug) where the crystal used to be. The second piece of evidence is BB-sized spheres of minerals scattered throughout the rocks at *Opportunity*'s landing site. These spherules are much like minerals that accumulate within porous, water-soaked rocks on Earth. A third piece of evidence comes in the form of crossbedding—a pattern that appears in rocks that can be attributed to the undulating motion of water across a sea floor. Crossbedding could also be caused by wind, but with all the other evidence, it seems highly unlikely that this pattern was formed by anything other than water.

With direct evidence of a long-standing salty sea on the Martian surface, this means that sometime in the past Mars was not only warm enough and had a thick enough atmosphere for liquid water to exist, but it was able to maintain that temperature for a long period of time. How long? Was it long enough for life to develop? Will we find evidence of that life? As you can see, we still have much more exploring to do before we can answer all these questions.

How many spacecraft have actually landed on Mars?
There have been five successful landings on Mars.

    ✺ NASA's *Viking 1* soft-landed on the planet—used a series of retro-rockets to slow its descent and gently touched down on the surface—on July 20, 1976.[32]

- NASA's *Viking 2* soft-landed on the planet on September 3, 1976.[33]
- NASA's *Pathfinder* bounced to a landing (encased in a protective cocoon of airbags) on July 4, 1997, with its *Sojourner* rover.[34]
- The Mars Exploration Rover *Spirit* bounced to a landing on January 4, 2004.[35]
- The Mars Exploration Rover *Opportunity* bounced to a landing on January 25, 2004.[36]

There have been many other missions to Mars—some succeeding, some failing. The Soviet Union's *Mars 3* lander, for example, touched down on the Martian surface on December 2, 1971, but it only survived for twenty seconds, transmitting static back to Earth before mysteriously shutting down.[37] Should this be considered a success or failure? It depends on how you look at it. (For more information, see the Missions to Mars section beginning on page 271.)

What would it be like to live on Mars?
*Cold, dry, and dusty, with very little air to breathe
(not that it would matter since what little air there is
doesn't contain enough oxygen to support human life).
However, since it has one-third the gravity of Earth,
you would be pretty light on your cold, dead feet.*

Before you could stand on the surface of Mars, you would have to find a way to overcome extremely cold temperatures and a very thin atmosphere. Even though it can get as hot as 70°F (20°C) during a summer day on the Martian equator, it can quickly drop below −100°F (−38°C) that same evening. Winter temperatures at the poles can drop as low as −220°F (−140°C), so be sure and pack your long johns.[38]

You'd better pack some life-support systems as well. The thin atmosphere that surrounds the planet is not really fit for human consumption. First of all, there is very little of it. At "sea level" on Mars, there is far less air than you would find at the top of Mount Everest. Even if the atmosphere were dense enough, it still wouldn't be any good for humans because it is made up almost entirely of carbon dioxide (the same as Venus, only much thinner).[39] There would not be nearly enough oxygen to support human life. In addition, the thin atmosphere allows dangerous amounts of ultraviolet radiation, x-rays, and gamma rays from the Sun to pass through to the surface—radiation that would broil delicate human skin. At night, the thin atmosphere does little to keep daytime heat from radiating back into space—hence the frigid nighttime temperatures.

If you'd managed to overcome these obstacles, the first thing you'd notice on the surface of Mars would be red. Red rocks, red dirt, red hills, red wherever you look. In addition, everything would be covered with a fine layer of red dust. If you happened to land during one of

Mars's frequent dust storms, even the sky would be reddish pink from all of the dust particles suspended in the atmosphere. Winds during these dust storms can gust up to sixty miles per hour (one hundred kilometers per hour), but have no fear. You wouldn't be blown away. With very little atmosphere for the winds to move around, these high-velocity gusts would have the same feel as a slight six-mile-per-hour (ten-kilometer-per-hour) breeze on Earth.[40] In addition, since sound can't travel very well through the thin atmosphere, you probably wouldn't even hear the dust storm approaching.

Besides being red in color, the type of landscape you would see would depend on where you are standing. If you happened to land within Valles Marineris, the largest canyon in the solar system, you would see a series of deep, intricate valleys with high rugged walls. If you landed on Olympus Mons, you probably wouldn't notice that you were on the largest volcano in the solar system. Instead, it would seem as if you were on a gently sloping hill that went on forever. If you touched down at the north or south poles, you would see frozen dust-covered ice caps made of water ice and carbon dioxide ice (dry ice). Scattered around the planet you would see other familiar sights such as craters, dried river beds, and smooth flat plains—all covered with reddish dust and rocks.

Although some of the features you would see on Mars were probably caused by the action of ancient waters (gullies, dried river beds, etc.), not a drop of liquid water exists on the surface today. With such cold temperatures and the extremely low atmospheric pressure, it would be a toss-up as to whether liquid water on the surface would evaporate before it froze. So any water that you'd bring to Mars would need to be kept in a sealed, insulated container.

From the dry, dusty Martian surface, the Sun would appear as a smaller, slightly fainter disk than what we are used to seeing on Earth. There would also be no large moon to brighten a Martian evening. While Mars has two moons, Phobos and Deimos, they are so small that they would appear only as bright starlike objects in the Martian sky.[41] From the planet, Phobos would appear as an extremely bright star whizzing across the sky three times a day. (This moon is so close to Mars that it only takes about seven hours and thirty-nine minutes to complete one orbit.) Deimos, on the other hand, wouldn't be as bright or as fast, taking thirty hours and seventeen minutes to complete one orbit.

Not only are the moons different, the length of the Martian day is different as well. Each day is twenty-four hours and thirty-nine minutes long, allowing for an extra nineteen and a half minutes of daylight and an additional nineteen and a half minutes at night to sleep (or not). While thirty-nine extra minutes a day doesn't seem like much, it does add up. Over the course of a week, that's an extra four and a half hours. It would take a while for your body to acclimate to the longer days. Mars also has a longer year. A Martian year is almost twice as long as Earth's, meaning that its seasons are twice as long as well. Longer summers are definitely a good thing. For most of us, longer winters are not.

Although humans on Mars would have many obstacles to overcome, they are not insurmountable. With the exception of Earth, the Red Planet is the least hostile of all the other planets in the solar system—for humans, that is. Mars may be our best bet if things go really bad on our home world. Just something to keep in mind.

What is the Great Red Spot on Jupiter?
A huge, rotating storm similar to a hurricane on Earth.

The Great Red Spot is one of the largest storms in the solar system. It was first seen in 1664 by the astronomer Robert Hooke, making it the first weather-related feature ever discovered on another planet.[42] Although the Great Red Spot has been storming on Jupiter ever since, it hasn't always had the same intensity. Over the years, this storm has varied in size from one to three times the size of Earth. (Here on Earth, we consider a storm as big if it covers one or two states. This storm would completely cover our entire planet—and then some!) Although it's called the Red Spot, the storm's color has varied over the centuries, ranging from a deep red color to a pale brown.

A close-up image of Jupiter's Great Red Spot can be found in the color insert section of this book.

Astronomers are not sure what caused the Great Red Spot to form, or what has kept it going all these years. One reason for its longevity may be a combination of Jupiter's fast rotation and the fact that different parts of the giant planet rotate at different speeds, a process called differential rotation. Even though Jupiter is more than eleven times larger than Earth, it rotates over twice as fast as our planet. In addition, the regions around the north and south poles of Jupiter rotate a little slower than the region around the equator. Gases near the poles take nine hours and fifty-five minutes to complete one rotation, while gases at the equator take nine hours and fifty minutes to spin around one time.[43] Although this rapid, uneven rotation causes turbulence and contributes to the huge storm, it does not explain what caused it to form in the first place, or how long it will last. Scientists are looking to Earth storms for some hints.

If the Great Red Spot follows the pattern of storm formation on Earth, it could be around indefinitely. Hurricanes on Earth form over large bodies of warm water. They gain strength from the warm water. When a hurricane passes over land, it leaves behind its source of energy. Consequently, the storm weakens and eventually dissipates. Jupiter has no land for the Great Red Spot to pass over, so this storm will not slow down like an earthly hurricane. It will take some other force to slow the Great Red Spot. We will have to wait and watch.

Could you land a spacecraft on Jupiter?
No.

Jupiter has no "surface" to land on. A spacecraft would sink through thicker and thicker clouds until the clouds were thicker than split pea soup. By then, the pressure would be so great that it would crush the spacecraft—which is exactly what happened to a small probe released into Jupiter's atmosphere on December 7, 1995. NASA's *Galileo Probe* plunged to a depth of 120 miles (200 kilometers) below Jupiter's upper cloud layer before it was crushed by pressure twenty-four times that of Earth's atmosphere. Just before its demise, it recorded winds of 400 miles per hour (650 kilometers per hour) and a temperature of 305°F (152°C).[44]

(For more details about trying to land on a gaseous body, see the explanation under the question "If the Sun is made of gas, could you fall right through it?" on page 35. Even though the Sun is much bigger than Jupiter, the idea is the same. The deeper you go into Jupiter, the greater the pressure.)

What would it be like to live on Jupiter?
Without any hard surface to land on, it would be next to impossible to live on this gas giant world—or any other gas giant planet, for that matter.

On a world made up almost entirely of gas, it may be possible for a specially designed spacecraft to hover within the clouds of Jupiter, much like a hot-air balloon does on Earth. However, conditions within the planet's upper atmosphere (including strong gravity, intense radiation, extremely high winds, turbulence, frigid temperatures, and powerful lightning, just to name a few) would almost guarantee that the stay would not be a pleasant one.

The upper atmosphere is really the only place you would have a chance to explore. As you would descend through the clouds, temperatures and pressures would quickly become unbearable. The *Galileo Probe*, which plunged into Jupiter's clouds on December 7, 1995, found that just 120 miles (200 kilometers) beneath the cloud tops, temperatures had increased from a frigid –166°F (–110°C) to a scorching 305°F (152°C) and the pressure had increased to twenty-four times what we are used to at sea level on Earth.[45] That was only 120 miles down. Jupiter's gaseous atmosphere extends down 40,436 miles (65,114 kilometers) to its dense rocky core. Pressures and temperatures at Jupiter's core would be incredibly high, beyond anything we could possibly survive. Bottom line—Jupiter and the other gas giants may be beautiful to look at from afar, but they would not be great places for humans to visit.

Are there oceans on some of Jupiter's moons?
Probably, but don't expect to take a swim in them any time soon.

Scientists believe that salty seas lie deep below the surfaces of three of Jupiter's largest moons. Europa's ocean seems to be fairly "close" to its surface, at a depth of 10 to 60 miles (16 to 100 kilometers).[46] Ganymede's ocean lies at a depth of 90 miles (150 kilometers).[47] Callisto's ocean lies about 120 miles (200 kilometers) beneath its surface.[48]

Since the oceans are so deep beneath the surface of each moon, it is currently impossible to have direct proof of their existence. However, observations of these moons taken by the *Galileo* spacecraft include data that are best explained by the existence of large bodies of salty water

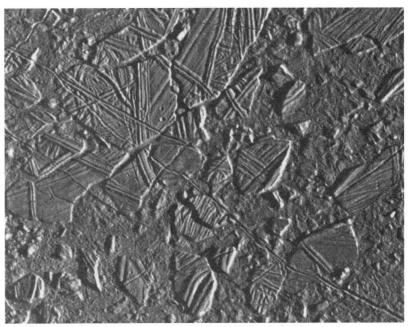

Broken chunks of ice on Europa resemble ice floes on Earth, suggesting the presence of liquid water or mushy ice near the surface. Image credit: NASA/JPL/University of Arizona

beneath the surface of each moon. What the spacecraft found were induced magnetic fields, magnetic fields that were not originally formed with the moons.[49] An induced magnetic field is caused by interactions with an external source. In the case of the Jovian moons, it is caused by the interaction of Jupiter's powerful magnetic field with some type of electrically conductive layer within each moon.

As each moon orbits the planet, the powerful influence of Jupiter's magnetic field stirs up, or induces, electrical currents within a conductive layer inside the moon. The faint electrical currents produce a magnetic field large enough to be detected by a spacecraft. The best candidate for the electrically conductive layer is liquid salt water.

What are the rings of Saturn made of?
Dust, ice, and rocks.

The beautiful ring system of Saturn consists of thousands of thin rings circling the planet. Within each ring are thousands of icy particles—particles that range in size from a grain of sand to the size of a house.[50] Even though the rings stretch out around the planet for thou-

A portion of
Saturn's intricate ring system.
Image credit: NASA/JPL/Space Science Institute

sands of miles, they are very thin, perhaps averaging 100 yards (91 meters) thick.[51] With literally millions of icy particles in orbit around Saturn, traveling through the rings could be like traveling through a blizzard.

Saturn's rings are kept in place by the planet's gravity and by the gravitational interactions of a few shepherd moons, small moons that orbit around Saturn within the outer rings. The shepherd moons help to herd the ring particles into position. Unusual ring features such as kinks, knots, and twists often result from the gravitational tug-of-war between Saturn, the shepherd moons, and the tiny ring particles.

The shepherd moons are also responsible for clearing certain gaps within the rings, such as the Cassini division and the Enke gap. The *Cassini* spacecraft (named after the astronomer Giovanni Domenico Cassini who discovered the Cassini division in 1675) flew through one of the larger gaps in the rings when it first arrived at Saturn in 2004.[52]

Is Saturn the only planet that has rings?
No.

All four of the gas giant planets (Jupiter, Saturn, Uranus, and Neptune) have rings, but Saturn's are by far the biggest and brightest.

Jupiter has one lone ring that consists of very dark material. The ring can't be seen from Earth. Astronomers didn't even know Jupiter had any rings when the first *Voyager* spacecraft flew by the planet. They took a chance and aimed the camera of the spacecraft where they thought any rings might be. As luck would have it, the camera did capture a picture of one lonely, dark ring.

Astronomers knew that Uranus had rings before *Voyager 2* flew by the planet. Not because they could see them, but because they couldn't see through them. Sound strange? Here's what happened. Astronomers were observing Uranus as it passed in front of a distant star. (This is called an occultation.) They expected to see the star disap-

Jupiter's ring. Image credit: NASA/JPL/Cornell

The rings of Uranus. Image credit: NASA/JPL

pear when Uranus passed in front of it. What they didn't expect to see was the star blinking on and off before Uranus got to it. After Uranus moved away from the star, the star blinked on and off again. Something surrounding Uranus was blocking the light from that star. That something turned out to be nine dark, narrow rings. When *Voyager 2* got to the planet in 1986, the spacecraft discovered two more rings, bringing the Uranian ring count up to eleven.[53]

Astronomers observed Neptune during an occultation to see if they could discover a ring system around that planet as well. What they found was confusing. Sometimes a star would blink on and off, and sometimes it wouldn't. They decided that Neptune must have ring segments instead of complete rings.

When *Voyager 2* sent back pictures of Neptune, they showed that the planet did indeed have complete rings—three of them, in fact. The rings were just too thin to block the light from a star. Scientists also found that the outer ring contained uneven clumps. The clumps were thick enough to block the starlight during an occultation and that was what astronomers on Earth had observed.

**Neptune's rings.
Notice the three brighter "clumps"
in the outer ring. Image credit: NASA/JPL**

Is Pluto really a planet?
*Officially, yes. However, there are some astronomers who want
to change Pluto's status from a planet to a Kuiper Belt object.*

The question regarding Pluto's status as a planet is a rather touchy one, which may never be fully resolved. There is as much emotion as logic involved with these arguments. And, as you can see below, most of the arguments used to support Pluto as a planet can also be used to argue that Pluto is a Kuiper Belt object—a small body composed of rock and ice that orbits beyond Neptune.

In support of Pluto's planet status:

1. **Its Orbit**—Pluto orbits the Sun, like the other eight planets.
2. **Its Size**—Even though Pluto is the smallest planet, it is by far the largest of all the other objects discovered beyond the orbit of Neptune.
3. **Its Moon**—Pluto has a moon,

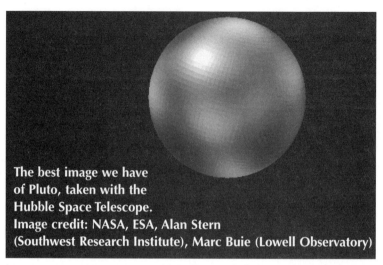

The best image we have of Pluto, taken with the Hubble Space Telescope. Image credit: NASA, ESA, Alan Stern (Southwest Research Institute), Marc Buie (Lowell Observatory)

like six of the other planets in the solar system. Only Mercury and Venus have no moons.

4. **An Atmosphere**—Pluto has an atmosphere. True, for most of its 248-year orbit that atmosphere is frozen on the ground. However, when Pluto approaches the Sun, its atmosphere of nitrogen, methane, and carbon monoxide thaws and surrounds the planet. Of all the planets, only Mercury doesn't have an atmosphere.

In support of Pluto's Kuiper Belt object (KBO) status:

1. **Its Orbit**—Pluto's orbit is tilted with respect to the Sun and other planets, meaning it travels above and below the orbital plane of the solar system (the ecliptic), much more so than any other planet. Pluto's orbit is also the most eccentric (egg-shaped) of all the planets. (The orbits of most planets are nearly circular.)
2. **Its Size**—Compared to the other eight planets in the solar system, Pluto is extremely small. In fact, even Earth's moon is larger than this tiny planet.
3. **Its Moon**—Pluto's moon, Charon, is not that large compared to the other moons of the solar system. However, when you compare the size of Charon to Pluto, the moon/planet ratio is very large. Below is a size comparison of planets with their largest moons.[54]

Our Moon	= 30% the size of Earth
Phobos	= 0.3% the size of Mars
Ganymede	= 4% the size of Jupiter
Titan	= 4% the size of Saturn
Titania	= 3% the size of Uranus
Triton	= 5% the size of Neptune
Charon	= 50% the size of Pluto

Since Charon is half the size of Pluto, some astronomers believe that the two should be considered either a double planet or a pair of Kuiper Belt objects.

4. **Its Composition**—The densities of Pluto and Charon indicate that the two are composed of a mixture of rock and ice, unlike any other planet.[55] This composition matches the composition of other recently discovered Kuiper Belt objects.

Eventually the Pluto planet issue will probably be decided by the International Astronomical Union (IAU), but even its decision—no matter what the decision is—will probably be contested. For now, Pluto is still officially a planet.

Is there life on any other planet in the solar system?
No, not that we know of.

For life as we know it to exist, certain requirements must be met: requirements such as temperatures that are not too hot or too cold, the presence of liquid water, and just the right amount of oxygen.

Spacecraft have flown by or landed on every planet in the solar system except Pluto. They have found no traces of life. Two spacecraft, *Viking 1* and *Viking 2*, took soil samples on the Martian surface and tested them for life-forms. The results were inconclusive.[56] While every planet in the solar system has its own unique beauty, Earth is the only one suited for life as we know it.

The planet Mercury has no atmosphere and its temperature varies from 660°F (350°C) during the day to –274°F (–170°C) at night. On Venus, you would have to survive an average temperature of 900°F (480°C) and an atmospheric pressure ninety times greater than Earth's. On Mars, the temperature isn't quite as harsh, with highs around 70°F (20°C) and lows around –220°F (–140°C). However, the Martian atmosphere is not the least bit friendly to humans. Its air is made almost entirely of carbon dioxide and its atmospheric pressure is less than one-hundredth of Earth's.

The gas giant planets—Jupiter, Saturn, Uranus, and Neptune—are made up mostly of hydrogen and helium gases and have no solid surface to stand on (although having a solid surface is not a criterion for life). Even though the gas giant planets are extremely cold at their cloud tops, temperatures and pressures increase rapidly as you plunge through the clouds. There could be a temperate zone that might support life somewhere within the cloud layers, but it would be unlike anything we have encountered so far. While Pluto has a surface to stand on, it is a very, very cold one. So cold, in fact, that for most of its 248-

year orbit, Pluto's thin atmosphere lies frozen on the ground. Living on these planets would be difficult, to say the least.

Though not a planet, some scientists believe that Europa, one of Jupiter's moons, may have a warm ocean beneath its crust that might be able to support life. If we consider that this ocean lies at least six miles (ten kilometers) beneath Europa's surface, any life that it may support would be extremely difficult for us to examine. Many kinds of life might survive under much harsher conditions than we have on Earth, but for now, the only life we have found in the solar system is right here.

Are there planets around other stars?
Yes.

Although very difficult to locate, astronomers believe they have discovered over one hundred extrasolar planets, planets orbiting other stars.[57] These planets are not Earth-like. Far from it, in fact. The extrasolar planets found so far are larger than Jupiter and orbit much closer to their stars.

Finding planets around other stars is very difficult, but not impossible. Have you ever had someone shine a flashlight in your eyes? It is difficult to see who is holding the flashlight because your eyes are blinded by the light. Now, instead of a flashlight, imagine one of those giant spotlights they use for movie premieres. And instead of a friend, imagine a small pea placed right next to the spotlight. Now, imagine the spotlight/pea combo about ten miles (sixteen kilometers) from you. Do you think you would be able to see the pea? This is the type of problem astronomers encounter when visually searching for extrasolar planets.

Because of the enormous distances involved and the extreme brightness of a star when compared to a planet, astronomers aren't able to see any extrasolar planets directly. So they have come up with two methods to "see" them indirectly: the transit method and the wobble method.

To help describe the transit method, let's go back to the spotlight/pea combo. Imagine the pea slowly moving across the face of the spotlight. As the pea transits (or moves in front of) the light, it blocks a pea-sized portion of the beam, thus causing the spotlight's overall brightness to decrease ever so slightly. Your eyes would have to be incredibly sensitive to detect the change, but it's there.

In the case of an extrasolar planet, astronomers carefully study the brightness of stars over a period of time, searching for minuscule changes that could indicate the transit of a planet across the face of a star.[58] If they detect a regular pattern of decreasing and increasing brightness, they could be "seeing" a planet in its predictable path, moving in front of its star one time each orbit. The problem with using this method is that the changes in brightness are extremely small and, as noted, extremely difficult to detect.

The second method for detecting extrasolar planets is known as the wobble method, or the radial velocity method.[59] Astronomers look for changes (wobbles) in the spectrum of the star to indicate the gravitational tug of a nearby planet. If there are no planets tugging on a star, its spectrum (the rainbow of colors that result when the light from the star is spread out into its individual colors) would remain constant. With a planet tugging on one side of the star, a minuscule shift toward the blue end of the star's spectrum would indicate the star was moving slightly toward us. (This shifting spectrum is also known as the Doppler shift.) Later, as the planet orbits to the other side of the star and its gravity tugs from the opposite direction, the star's spectrum would shift toward the red, meaning that the star is now moving slightly away from us.

The wobble method is currently the most popular method with which to detect extrasolar planets. However, it, too, has its problems. Only the largest, most massive planets provide enough of a gravitational tug to cause a detectable shift in a star's spectrum. And even then, the shifts are very slight and difficult to measure. In addition, this method merely indicates the presence of something tugging on the star. There is no way to determine the mass of that object. Without knowing an object's mass, it's hard to tell if it's a planet, a brown dwarf, or even another small star.

So for now, while there appear to be planets orbiting around other stars, the vast distances of space will limit us to mere hints of their existence.

ABOUT ASTEROIDS, COMETS, METEORS, AND OTHER SPACE JUNK

What is an asteroid?

What is the asteroid belt?

What is the largest asteroid?

Is it dangerous to travel through the asteroid belt?

Are all asteroids found within the asteroid belt?

What formed the asteroid belt?

Did an asteroid really wipe out the dinosaurs?

What is a comet?

How big is a comet?

Where do comets come from?

Do comets move quickly across the sky?

Why is Halley's comet so important?

When is the next time we will see a bright comet?

What is a meteor?

What is the difference between a meteor and a meteorite?

What are meteors/meteorites made of?

What is the difference between a "falling star," a "shooting star," and a meteor?

Why do meteors leave a streak of light across the sky?

What is a meteor shower?

When is the next meteor shower?

Where do meteor showers get their names?

What are Near Earth Objects (NEOs)?

What are the Centaurs?

What is an asteroid?
A rock made of stone, iron, or a combination of both that travels around the Sun.

Asteroids come in many different shapes and sizes. Mostly they look like what they are—chunks of rock and metal.

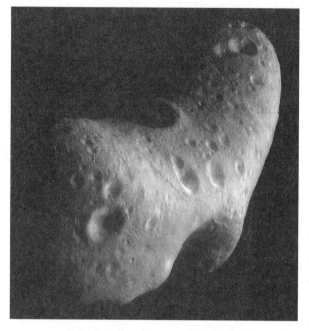

Asteroid 433 Eros orbits the Sun once every 1.76 years. It is 21 miles (33 kilometers) long, 8 miles (13 kilometers) wide, and 8 miles (13 kilometers) thick. This image was taken by the *NEAR Shoemaker* spacecraft in orbit around the asteroid. Image credit: NASA/JHU-APL

Asteroids are sometimes called *minor planets*. The "minor" part of the name is because they are much smaller than the planets we know. Asteroids range in size from less than half a mile (about one kilometer) to almost 600 miles (1,000 kilometers) in diameter.[1] In comparison, Pluto, the smallest planet, is 1,430 miles (2,300 kilometers) in diameter.

The "planet" part of the name comes from the fact that asteroids orbit around the Sun. We know that Earth travels around the Sun once every 365.25 days, or one year. The asteroid Vesta travels around the Sun once every 3.63 years.[2]

What is the asteroid belt?
An area between the orbits of Mars and Jupiter where as many as a million asteroids can be found.

Based on readings taken by the European Space Agency's *Infrared Space Observatory*, astronomers estimate there are between 1.1 and 1.9 million asteroids larger than a half mile (one kilometer) in size between the orbits of Mars and Jupiter.[3] This number does not take into account any asteroids smaller than a half mile, so the numbers are probably much larger. Of these millions, astronomers have officially cataloged over 40,000 asteroids.[4]

Even though there are so many of them, asteroids are not easily seen because they are so small. Of the 40,000 cataloged asteroids, only about 230 are larger than sixty miles (one hundred kilometers) in diameter.[5] Most of the asteroids are much smaller.

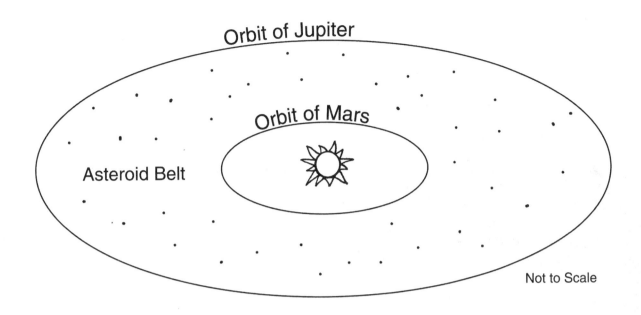

What is the largest asteroid?
Ceres.

The asteroid Ceres is a slightly flattened sphere whose diameter ranges from 577 to 602 miles (930 to 970 km).[6] It lies at a distance of 2.77 AUs from the Sun, and takes 4.6 years to complete one orbit.[7] Its orbit places Ceres between the planets Mars and Jupiter, making it the largest member of the asteroid belt.

When Ceres was first discovered on January 1, 1801, astronomer Guiseppe Piazzi thought he had discovered the eighth planet of the solar system.[8] (Uranus, the seventh planet, had only just been discovered twenty years earlier.) His excitement faded over time, however, when further observations indicated that his new discovery was really too small to be considered a planet. But what exactly was it? It took astronomers fifty years and the discovery of several other similar objects before they were able to figure out what these minor planets, or asteroids, were and how they fit within the solar system.[9]

Ceres is by far the largest asteroid, which explains why it was discovered first. Only two asteroids, Pallas and Juno, come close to matching Ceres in size. Pallas (the second asteroid discovered) and Juno (the third asteroid discovered) are both about 180 miles (300 kilometers) in diameter.[10]

Astronomers have charted about 225 other asteroids that are larger than sixty miles (one hundred kilometers) across. The vast majority of the other cataloged asteroids are much smaller, on the order of half a mile (one kilometer) across.

Is it dangerous to travel through the asteroid belt?
Not really.

When scientists sent the first spacecraft (*Pioneer 10*) to visit Jupiter, they were concerned about the asteroid belt. They didn't know if a spacecraft could travel through this crowded portion of space and survive, or if it would be pulverized by an asteroid. Unfortunately, there was no way to avoid passing through the belt.

When *Pioneer 10* entered the asteroid belt, nothing happened.[11] The spacecraft—which had instruments onboard to record impacts—detected no increase in the number of hits as it traveled through the belt. And nothing has happened to any other spacecraft passing through the asteroid belt since then. It was as if the asteroid belt didn't exist. Is this strange? Not really.

Even though there may be over a million asteroids in the asteroid belt, there is still quite a bit of empty space within this region. The belt between Mars and Jupiter is approximately 140 million miles (223 million kilometers) wide. The asteroids orbit within this belt in an area covering 265 trillion square miles (442 trillion square kilometers).[12] Even with one million asteroids in the belt, this leaves each asteroid with an average of 2.6 billion square miles (4.4 billion square kilometers) all to itself. With so much space between them, the asteroids are relatively easy to avoid. In fact, both the *Galileo* and the *Cassini* spacecrafts had to change their flight paths in order to fly by medium-sized asteroids. In other words, the asteroid belt is not such a dangerous part of space to fly through after all.

Are all asteroids found within the asteroid belt?
No.

The asteroids found within the asteroid belt are called "main-belt" asteroids. At least four other types of asteroids can be found in other parts of the solar system, including Trojan asteroids, Amor asteroids, Apollo asteroids, and Aten asteroids.[13]

Trojan asteroids can be found sharing the orbits of Jupiter and Neptune.[14] These asteroids actually orbit along the same path as the larger planet, either slightly in front of the planet or slightly behind it. These asteroids are trapped in what are known as Lagrange points. When two bodies (such as a star and a planet) orbit around each other, there are points within the orbit where the centrifugal and gravitational forces balance

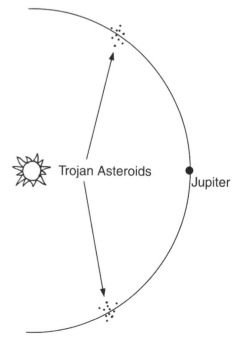

each other out. These points are known as the Lagrange points, named after the mathematician Joseph Louis Lagrange who first predicted their existence in 1772.[15] Astronomers discovered the first Trojan asteroids in orbit around Jupiter in 1906. A Trojan asteroid around Neptune was discovered only recently. There may be Trojan asteroids around the other planets (including Earth), but none has been discovered to date.

Other asteroids have very elliptical, or egg-shaped, orbits that bring them into the inner solar system. The Amor asteroids cross the orbit of Mars, the Apollo asteroids cross the orbit of Earth, and the Aten asteroids cross the orbit of Venus. The Apollo and Aten asteroids are also known as Near Earth Objects, or NEOs, for an obvious reason—they are near Earth.

What formed the asteroid belt?
Astronomers believe that the asteroid belt formed along with the original solar system, but that gravitational interactions at the belt's location didn't allow the material to coalesce into a larger body.

If Jupiter hadn't been so large and so close to this region of space, all the matter orbiting in what we know now as the asteroid belt might have come together to form a small planet. However, Jupiter's strong gravity kept tugging on all of the small pieces, attracting some of the material to the large planet, flinging some matter completely out of the solar system and never allowing the remaining pieces to come together long enough to stick. The result of all this gravitational interference is the many asteroids of the asteroid belt.

For a long time, astronomers thought that the asteroid belt was the result of a planet that had broken apart or shattered. However, if you put all of the asteroids together, the resulting planet would be smaller than Pluto. This "planet" doesn't fit in with the rest of the solar system—and it brings up another puzzling question: What could cause a planet to break apart? The forces required to destroy a planet are almost beyond comprehension. It would take an incredibly catastrophic event such as a collision with an object of comparable size or the passage of a massive nearby star. An event of this magnitude would have residual effects throughout the solar system that should be visible even today. However, with the exception of the asteroids within the asteroid belt, we don't really see anything to suggest such a calamity occurred. By now, most astronomers have given up on the idea of a shattered planet and agree that asteroids are simply leftover debris from the formation of our solar system.

Did an asteroid really wipe out the dinosaurs?

Yes and no. While some may have died during the actual impact, most dinosaurs were not killed by getting knocked in the head by a space rock! They died soon afterward, though, because of the dramatic changes in climate caused by the impact.

Scientists can never be exactly sure what happened to the dinosaurs. However, they have gathered a great deal of credible evidence to support the theory that an asteroid impact sixty-five million years ago contributed greatly to the demise of the large beasts. While the actual impact would have killed any dinosaur that was unlucky enough to be living within a couple hundred miles of the impact site, the vast majority of dinosaurs were not killed by the impact but by the resulting drastic changes in climate.

According to theory, an asteroid six miles (ten kilometers) in size slammed into Earth sixty-five million years ago in the area we now know as the Yucatan Peninsula in Mexico. Tremendous heat from the impact instantly vaporized enormous amounts of water and debris, flinging it high into the upper atmosphere. A searing firestorm and damaging shock wave traveled out from the impact site, destroying everything for hundreds of miles in every direction. The effects from this tremendous impact were soon felt all over the planet as high-level winds carried the dust and debris around the globe. The debris in the air kept sunlight from reaching the surface, and temperatures began to drop worldwide. The cold, dark conditions lasted for many months, if not years.[16]

The lack of sunlight and the cold temperatures caused many plants to die. The herbivores (vegetarian dinosaurs) began to die off from the lack of food. Carnivores (the meat-eating dinosaurs) found their food supply—the herbivores—dying off and they soon followed. It took several years for the atmosphere to cleanse itself of all the impact debris and even longer for the climate to return to normal. By then, however, it was too late for the dinosaurs.

This dinosaur-killing asteroid theory was first introduced in the late 1970s when it was discovered that a high concentration of the element iridium existed in the geological boundary marking the end of the Cretaceous period (the time when dinosaurs reigned) and the beginning of the Tertiary period (when mammals took over). Iridium is normally an extremely rare element on Earth, yet at different locations around the globe, geologists were finding large concentrations of the element within this one boundary layer.

Luis Alverez (a physicist) and his son Walter (a geologist) were the first to suggest that this unusual concentration of iridium came from an asteroid impact.[17] Asteroids are known to contain large quantities of iridium. The impact of a large, iridium-rich asteroid would destroy the space rock, and the resulting clouds of debris would spread a fine layer of iridium-laden dust around the globe. To account for the levels of iridium they were seeing in this one layer, the Alverezes calculated that the original asteroid would have been approximately six miles (ten kilometers) in size.

While the theory made sense, no one had found any evidence of the large crater that should have resulted from such an impact. Then, in the early 1990s, geologists found evidence of a large crater centered just off the northern coast of the Yucatan Peninsula in Mexico, near the small coastal town of Chicxulub.[18] Radioactive dating indicates that this 112-mile (180-kilometer) crater, dubbed the Chicxulub Crater, was formed approximately sixty-five million years ago. Many believe this crater is the "smoking gun" that points to the fate of the dinosaurs. However, there are competing theories that suggest "Earth-based" causes (such as volcanism or disease) may have sealed the fate of the dinosaurs. The research continues.

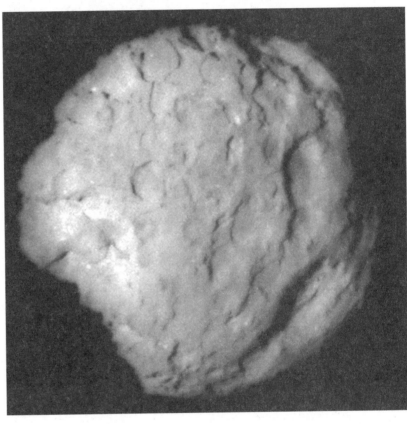

A close-up of comet Wild 2 shows the nucleus of a comet. Image credit: NASA/JPL

What is a comet?
A dirty snowball (a few miles in diameter) that orbits around the Sun.

A comet is a large, dirty snowball that spends most of its time in the far reaches of our solar system. Millions of miles away and very dark, a distant comet is almost impossible to see. However, when its orbit brings it close to the Sun, the comet begins to brighten as things literally heat up. Portions of the surface of the dark, dirty snowball—called a nucleus—begin to vaporize. The resulting cloud of gas and dust that is released is called a coma. The coma surrounds the nucleus and hides it from view. Solar wind and radiation pressure push against the material in the coma, blowing it away from the nucleus and forming two graceful tails.

The two types of tails associated with a comet are the dust tail and the ion tail.

The dust tail is the brightest and most easily visible of the two. It is composed of the solid debris—mainly dust particles—released from a comet when it melts. A combination of solar wind (a stream of charged particles flowing away from the Sun) and solar radiation pressure (reflected and absorbed photons of sunlight exerting pressure on the debris) push this debris away from the coma and form the dust tail.[19] In addition, since these solid particles within the dust tail have some mass to them, they are also affected by the Sun's gravity. As the solar

Parts of a comet ...

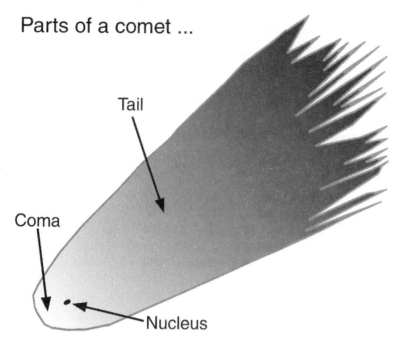

Tail

Coma

Nucleus

gravity tugs on each individual piece, the overall tail develops a slight curve. The dust tail is illuminated by reflected sunlight and therefore appears white in color.

The ion tail is composed of gases (ionized atoms and molecules) released from the nucleus as the ice melts.[20] The extremely lightweight ions are at the mercy of the solar wind. Thus, the ion tail always points directly away from the Sun as if blown in that direction by a strong gust. The ion tail is bluish in color from the glowing ionized atoms.

Both the white dust tail and the blue ion tail of a comet can be seen in the color insert section of this book.

The most unusual feature of the comet's tail is that it is not always behind the comet. Since particles within the tail are blown around by the Sun, the tail is always pointing away from the Sun, no matter which direction the nucleus is traveling. In other words, a comet's tail can either be streaming behind it (as the comet is approaching the Sun) or it can be ahead of the nucleus (as the comet is moving away from the Sun).

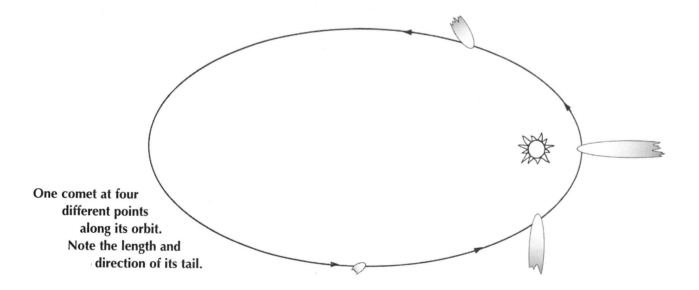

One comet at four different points along its orbit. Note the length and direction of its tail.

How big is a comet?
A few miles to several million miles across,
depending on which part of the comet you are talking about.

A comet's nucleus is a few miles across (the size of a city). Its coma can be thousands of miles across (the size of a planet). Its tail can stretch for millions of miles through space (rivaling the distance between planets).

Where do comets come from?
The Kuiper Belt and the Oort Cloud—
comet reservoirs in the far reaches of our solar system.

Astronomers believe that short-period comets (comets that take less than two hundred years to travel around the Sun) come from a region of space known as the Kuiper Belt.[21] Medium- and long-period comets (comets that take hundreds or thousands of years to complete one orbit) come from the more distant Oort Cloud.[22]

The Kuiper Belt begins near the orbit of Pluto and is believed to stretch outward for 500 AUs.[23] It is approximately 10 AUs thick.[24] (Recall that one astronomical unit [AU] is equal to 93 million miles or 150 million kilometers.) The Kuiper Belt is home to thousands of small icy bodies that were created here during the formation of our solar system. At this great distance from the Sun, the original solar nebula was too thin to coalesce into large planets. Instead, it formed large numbers of small icy bodies.

These small comet bodies usually orbit peacefully, minding their own business. Occasionally, however, a collision between bodies or a tug of gravity from a close flyby of Neptune disturbs the area and sends some along a new path.[25] Each new orbit could take a potential comet completely out of the solar system or in toward the Sun. The new comets that head in toward the Sun still orbit roughly within the plane of the ecliptic (the plane in which most planets revolve around the Sun). While astronomers have identified many short-period comets, they usually don't make headlines because their appearance is rather faint. Since they have been coming close to the Sun every few years, most of the bright ices have already melted from their surfaces.[26]

Medium- and long-period comets are a different matter, however. These comets originally formed in the region of space between Jupiter and Neptune. Caught in the gravitational tug-of-war between the larger planets, these small icy bodies were soon thrown out into the far reaches of the solar system, unable to hold their own orbits against the "big boys." Their new orbits were tweaked even farther by the passage of a nearby star. Eventually, the solar system was completely encircled by a sphere of small icy bodies. This sphere, which lies at a distance of 50,000 AUs from the Sun, is the Oort Cloud. Occasionally, the orbits of some of these icy bodies are disturbed, and they begin their journey toward the inner solar system, a journey that can take hundreds to thousands of years.

Medium- and long-period comets are usually much brighter than short-period comets, simply because they haven't been around the Sun as much. Since they heat up only once every few hundred or thousand years, it takes a much longer time for all of their bright ices to melt away. Because these comets originate from a sphere of material that surrounds the solar system, they can come at us from any angle or inclination. Their orbits completely ignore the plane of the ecliptic.

Do comets move quickly across the sky?
No.

Over the course of a long evening, you may see that a comet has moved very slightly when compared to background stars, but a comet will not blaze across the sky in a matter of moments like meteors.

Some comets do move very quickly along in their orbits, but instead of a streak of light flashing across the sky, all you would be able to see from night to night is a slight change in position with respect to the background stars. The reason for this relatively slow pace across our sky is because comets are very far away from us. The farther away an object is, the slower it appears to be moving. For an Earthly example, imagine a jet flying overhead. The jet may be traveling five hundred miles per hour (eight hundred kilometers per hour), yet from where you stand on the ground, it may look as though it is barely moving. A car passing in front of you at a speed of sixty miles an hour (one hundred kilometers per hour) will appear to be moving much faster than the jet simply because it is much closer to you.

Why is Halley's comet so important?
Halley's comet helped astronomers realize exactly
what comets were and how they fit into our solar system.

In the past, no one understood exactly what comets were. They moved across the sky like planets but at unpredictable times and from strange directions. They appeared out of nowhere, grew in size, developed strange tails, then faded to nothing. Many of our superstitious fore-bears believed that comets were omens of misfortune. Some took them as signs from the gods. It is said that William the Conqueror took the appearance of a comet in the year 1066 as a sign that it was time to invade England.[27] (This could have been interpreted as a good sign—if you were French—or a bad sign—if you were English.)

After the appearance of an extremely bright comet in 1682, a scientist named Edmund Halley decided he would try to calculate the comet's path. In doing his research, he noted that there had been several bright comets throughout history, some that had even passed through the same region of the sky as the one in 1682 (including the one seen by William the Conqueror). After many years of calculating, he concluded that many of these different sightings (which took place every seventy-five to seventy-six years) were not different comets, but the same comet returning over and over again.[28] Halley predicted that the comet would appear again in the year 1758. Unfortunately, he didn't live long enough to see that his prediction was correct. The returning comet was named in his honor.

In the past, Halley's comet has been a spectacular sight in the night sky. However, the last time it approached the Sun, in 1986, Earth was not in a good position to observe the comet. With all the publicity from the news media, many people were excited about its return but were disappointed when all they saw was a faint, fuzzy blob. Astronomers, however, were able to get a good view of it using the European Space Agency's *Giotto* spacecraft. *Giotto* flew within 370 miles (596 kilometers) of Halley's comet on March 13, 1986, and captured the first close-up image of a comet's nucleus—a dirty snowball five by five by ten miles (eight by eight by sixteen kilometers) in size.[29]

The next time Halley's comet makes an appearance in the inner solar system, in the year 2061, the comet will be closer to Earth and promises to be a beautiful sight.

The nucleus of Halley's comet, as seen by the *Giotto* spacecraft. Image credit: ESA

When is the next time we will see a bright comet?
Hard to say. It depends on when nature sends a bright one our way.

In the year 2061, astronomers are certain that there will be a bright comet in our skies—Halley's comet. In all likelihood, there will be other bright comets between now and then, but astronomers can't predict exactly when they will appear because these comets haven't been discovered yet.

Comets are predictable—once they have been observed and their orbits calculated. Astronomers know the orbits of many comets. However, most of the comets that have been identified are either fairly faint, or rather distant, or what astronomers call long-period comets—they pass close to the Sun once every few thousand years. The next bright comet will most likely be a long-period comet that has yet to be discovered.

Since 1996, there have been two bright comets that have lit up our skies: comet Hyakutake and comet Hale-Bopp. Both comets were discovered just before their close approach to the Sun and both put on a brilliant show in the night sky. In 1996, comet Hyakutake was discovered by amateur astronomer Yuji Hyakutake only two and a half months before its close approach to Earth.[30] Astronomers calculated the comet's orbit and found that it had been 8,000 years since this particular comet had visited the inner solar system and it would be another 14,000 years before it would return.[31]

Comet Hale-Bopp. Notice the incredible detail within the dust tail (pointing up and to the right). The fainter, more diffuse ion tail can be seen in the lower half of the image. Image credit: Lowell Observatory

Comet Hale-Bopp was so big and bright that it was discovered by Alan Hale and Thomas Bopp while still well beyond the orbit of Jupiter. Astronomers were able to track Hale-Bopp for over a year and a half before it made its close approach in March 1997. Astronomers calculate that it had been 4,200 years since this comet had visited the inner solar system and would be another 2,380 years before it would return.[32]

So for now, Halley's comet in 2061 is still the best bet for the next bright comet. We hope that we won't have to wait that long, but only nature knows what's coming our way.

What is a meteor?
A piece of dust or rock that burns up as it enters our atmosphere.

A meteor starts out as a piece of rock or dust floating in outer space. As Earth travels around the Sun, it sometimes runs into these pebble-sized rocks and dust particles. When they hit Earth's atmosphere, the rocks and dust burn up, leaving a bright streak across the sky. Most of the time, they burn up completely. Other times, some of the rock survives to land on Earth.

What is the difference between a meteor and a meteorite?
One is up in the air and one is on the ground—the only difference is location.

A meteor is a rock traveling through Earth's atmosphere. If that same rock makes it all the way down to the ground, it becomes a meteorite.

What are meteors/meteorites made of?
Stone and iron.

There are three major types of meteors/meteorites: stony, iron, and stony-iron.

Stony-type meteors make up about 92 percent of all the meteors.[33] If you see a meteor in the sky and watch it hit the ground, chances are it will be a stony meteorite. However, if you were to stumble upon a stony meteorite lying on the ground, you probably wouldn't notice that it was a space rock. Stony meteorites are similar in composition and weight to Earth rocks. As a consequence, they blend in with Earth rocks and are easy to overlook.

If you stumbled across the second type of meteorite, an iron meteorite, you would probably know something was strange. Iron meteorites make up only 6 percent of the meteor population, yet they are the most commonly found meteorites.[34] Since they are composed of 80 to 90 percent iron and 10 to 20 percent nickel, they are very heavy for their size! If you tried to pick up one of them, you would immediately notice its unusual weight. It would also attract a magnet, something most Earth rocks don't do.

The third type of meteorite is a stony-iron meteorite, which is made up of a combination of stone and iron. This type of meteorite is the rarest, making up only 2 percent of the meteor population.[35]

What is the difference between a "falling star," a "shooting star," and a meteor?
Nothing.

They are all different names for the same thing: a piece of rock or dust that burns up as it enters our atmosphere. They have nothing to do with the stars we see at night. Instead, they are all the result of nearby space debris hitting Earth's atmosphere.

Why do meteors leave a streak of light across the sky?
Friction between a meteor and Earth's atmosphere causes the space rock to heat up rapidly and burn as it passes through the air.

A meteor hits Earth's atmosphere at such a high speed that it is just like slamming into a hard wall. The meteor pushes through the dense atmosphere, overcoming the friction of the thick air at the cost of heat. Consequently, the space rock heats up to the point where it begins to burn. The trail of light you see in the night sky is the meteor as it burns its way through the atmosphere. Sometimes meteors are large enough that they don't burn up completely as they pass through the atmosphere. The surviving rock strikes the ground and forms a crater.

What is a meteor shower?
When you can see at least five meteors in an hour.

On any given night, you can see three or four meteors every hour—if you happen to be looking in the right places at the right times. That averages out to one meteor every fifteen or twenty minutes. During a typical meteor shower, you can see anywhere from five to one hundred meteors per hour. At that rate, you can expect to see one meteor every twelve minutes to one every forty-five seconds.

A meteor shower occurs when Earth passes through a dusty part of space. The dust is typically leftover debris from the passage of a comet. When our planet travels through these comet trails, the comet itself is usually long gone. All that remains are the dust, rocks, and other debris that were released when a portion of the comet's surface melted. When this debris hits our atmosphere, it heats up and a meteor is born. Lots of debris means lots of meteors, and all of a sudden you have a shower.

Every year our planet passes through the same dusty trails at about the same time, allowing astronomers to predict when meteor showers will occur. What astronomers can't predict is just how dusty a comet trail will be at any given time. Since comets are an uneven mixture of ice and dust, sometimes a large cloud of dust melts away from the comet, and sometimes it doesn't. Some comets simply contain more dust than others. And depending on how long it has been since the comet passed along the trail, the debris field may have thinned out as the dust and rocks slowly move away from each other. All of these factors work together to make it difficult for astronomers to predict the intensity of each shower.

Sometimes astronomers get it right, though. For the Leonid meteor shower of 2001 and 2002, astronomers took into account the recent passage of the comet Swift-Tuttle and believed that there should be fresh batches of dust and debris concentrated in the specific area that Earth would be passing through.[36] Their predictions were correct. During a few brief hours, there were literally thousands of meteors bombarding our planet, giving rise to a new term, the meteor storm.

When is the next meteor shower?
Listed below are some of the bigger showers that occur throughout the year.

Included in the list are the names, dates, when the peak of the shower occurs (when you are likely to see the most meteors), and about how many meteors you can expect to see per hour. These showers can be seen from all over the world. Keep in mind that the numbers listed under hourly rates are averages and may vary from year to year.[37]

Annual Meteor Showers

Shower Name	Dates	Peak Date	Hourly Rate
Quadrantid	January 1–5	January 3	40
Lyrid	April 20–24	April 22	15
η (Eta) Aquarid	May 3–6	May 5	20
δ (Delta) Aquarid	July 29–31	July 30	35
Perseid	August 10–14	August 12	50
Orionid	October 20–22	October 21	30
Taurid	November 3–6	November 5	15
Leonid	November 13–19	November 17	15
Geminid	December 10–15	December 13	50
Ursid	December 21–23	December 22	15

Note: The Moon is a big factor when it comes to observing meteors. If the Moon is above the horizon during the peak of a shower, its light will drown out all but the brightest meteors. The best time to view a meteor shower is during its peak night when there is no Moon in the sky (when the Moon is near its New Moon phase). The weather section of your local paper should have a listing of dates of the various Moon phases. (For other meteor shower viewing tips, see p. 211 of the Quick Facts section.)

Where do meteor showers get their names?
From the constellations they appear to come from.

Debris left behind by a comet gradually drifts apart over a long period of time, but most of it remains concentrated along the comet's original path. At the point where Earth intercepts that path, you see the highest number of meteors. From our vantage point on the surface, those meteors appear to be coming from one specific part of the sky, the area of space in which the comet passed. Whatever constellation of stars happens to lie behind the comet's trail at that point gives the meteor shower its name.

For example, most of the meteors you would see during the Leonid meteor shower appear to originate in the constellation of Leo, the Lion.

What are Near Earth Objects (NEOs)?
*Any comet, asteroid, or other space debris whose orbit brings it
close to Earth or actually crosses Earth's orbit.*

There are currently approximately 2,700 known Near Earth Objects (NEOs) with more being discovered every year. Almost 600 of these NEOs are designated as potentially hazardous asteroids (PHAs), meaning their orbits pose an impact threat to Earth.[38] None of these potentially hazardous asteroids are on a collision course with our planet, at least not for the next few hundred years. But there may be others—as yet undiscovered—that are on target with our planet. That is why astronomers are watching the skies, looking for NEOs.

There have been close encounters with NEOs throughout Earth's history. In 1908 a small asteroid about 280 feet (80 meters) in size exploded in the atmosphere over the Tunguska region of Siberia. The explosion knocked down trees for thirty miles (fifty kilometers) in all directions and was heard over six hundred miles (one thousand kilometers) away.[39] Sixty-five million years ago, an asteroid six miles (ten kilometers) in size smashed into what is now the Yucatan Peninsula in Mexico. The resulting drastic changes in climate led to the extinction of the dinosaurs.

Astronomers have determined that if an object a half mile (one kilometer) in size or larger were to hit Earth, serious damage would result. In response to the impact threat, observatories around the world have initiated programs designed to find and track NEOs. The goal is to be able to predict when and if an object is on a collision course with Earth and what to do just in case one is headed our way. With enough warning, we may somehow be able to save our planet.

What are the Centaurs?
*Small icy bodies that circle the Sun in unstable orbits
between Saturn and Neptune.*

In Greek mythology, the Centaurs were a rowdy race of beings that had the head and torso of a human and the body and four legs of a horse.[40] In astronomy, Centaurs are proving to be mysterious and perhaps a bit rowdy themselves. They are certainly giving astronomers a run for their money when it comes to explaining their existence and composition.

The first Centaur, Chiron, was discovered on November 1, 1977, by Charles Kowal.[41] Astronomers were puzzled by its unusual orbit, which took it across the path of Saturn and then back out almost as far as Uranus. The gravity of these two planets will not allow Chiron to maintain this Saturn-crossing orbit for long. Sooner or later, a close encounter with Saturn will fling Chiron off its current path and on to parts unknown. Because of this unstable, short-lived orbit, it appears that Chiron is merely a temporary visitor to this region of space. Where Chiron came from and what exactly this strange body is are two questions that astronomers are still trying to answer.

At first, they believed that Chiron was a very large comet. Jets of material had been observed streaming away from its surface, and the hint of a coma was seen beginning to form around its body. However, the activity soon stopped, and Chiron took on all the characteristics of a typical asteroid (a dead space rock with no activity whatsoever). Then the activity began again.[42] It soon became obvious that these outbursts were not tied to periods when Chiron was close to the Sun—which is unlike a typical comet. In addition, these bursts of activity never lasted for very long.

Astronomers puzzled over the one, lone Centaur for fifteen years. Then in 1992 a second Centaur was discovered. Given the name Pholus, its discovery just confused the issue even more. While its unstable orbit was similar to Chiron's, Pholus was much redder in appearance, signifying a different compositional makeup.[43] If the two Centaurs came from the same place, one would think that they would be made up of the same type of materials. Several more Centaurs have been discovered in the last few years. While all of their orbits are unstable, with some stretching as far out as Neptune, they are all very different in their physical attributes.

In an attempt to explain where these Centaurs came from, astronomers turned to the Kuiper Belt. They are suggesting that Centaurs may be former Kuiper Belt objects that have taken a detour on their way to becoming comets of the inner solar system. When an object from the Kuiper Belt is disturbed and begins its long trek toward the inner solar system, it must first cross over the orbits of the large outer planets. It is entirely possible that some of the small bodies get caught up by the gravity of one of the larger planets. These Centaurs could remain in an unstable orbit for a while, until a gravitational encounter with Saturn or Uranus flings them elsewhere in the solar system.

While this sounds like a good theory, there remain a few unanswered questions. For example, Chiron is about 125 miles (200 kilometers) in size, making it smaller than most Kuiper Belt objects that have been discovered to date, but larger than any comet ever observed.[44] Perhaps during their brief stay in these unstable orbits, the Centaurs lost large amounts of material from these strange outbursts. So by the time they were ejected from the unstable orbit and allowed to continue on toward the inner solar system, they were more along the size of a typical comet. That may explain the differences in size, but the difference in composition is still in question.

It will take much more study before astronomers understand the true nature of these mysterious entities. All they know for sure about Centaurs is that they are temporary citizens of the outer solar system. Where they came from and where they will end up is literally still up in the air.

ABOUT EARTH

Why does learning about astronomy help us understand Earth better?
Why is Earth the only planet with liquid water?
What is the aurora borealis, popularly known as the northern lights?
What is the ozone layer and what's wrong with it?
Why does the Sun move across the sky?
What causes the seasons?
Why is it cold in winter and hot in summer?
If it's winter in North America, why is it summer in Australia?
What is the winter solstice?
What is the summer solstice?
What is the vernal equinox?
What is the autumnal equinox?

Why does learning about astronomy help us understand Earth better?
Earth is part of the solar system.

**Earth as a planet.
Image credit: NASA**

Astronomy is the study of space and everything found within it—planets, moons, stars, nebulae, black holes, comets, asteroids, and much more. The planet Earth, along with its moon, is one of nine planets revolving around a star. Our Sun, as well as all the stars we see in the night sky, lies within one spiral arm of the Milky Way, our home galaxy. The Milky Way has at least four companion galaxies hovering close by. These five galaxies share a region of space with almost forty other galaxies, forming what is called the Local Group.[1] Beyond our local group of galaxies lie thousands of superclusters of galaxies.

Our tiny planet orbiting an average star is a small but integral part of the universe in which we live. While the study of astronomy is fascinating as it examines foreign worlds and distant stars, astronomers are also learning about Earth in the process. For example, by comparing ancient volcanoes on Mars with the volcanoes in Hawaii, astronomers learn more about both. By observing impact craters and debris patterns on the Moon, Mercury, and Mars, astronomers simultaneously learn more about these other planets and can also better identify impact regions on Earth. By studying the thick, soupy atmosphere of Venus, astronomers can gain insight into its atmosphere and that of primeval Earth and the dangers facing Earth's atmosphere today, such as the ozone hole and the greenhouse effect.

In addition to the planets, astronomers study the Sun in an attempt to learn what triggers variations in its energy output. After all, small changes in the Sun's surface temperature could cause the next ice age or worldwide drought here on Earth. Increased activity on the Sun's surface, such as solar flares or coronal mass ejections, can disrupt radio and satellite communications all around the globe.

Besides our own star, astronomers study similar Sun-like stars. Observing stars at different stages of their lives allows astronomers to piece together the different phases our Sun will undergo during its lifetime and give us humans an idea of what's to come. So you can see, all the knowledge gained by the study of astronomy helps in our understanding of our own planet Earth.

Why is Earth the only planet with liquid water?
It all has to do with location, location, location.
Earth finds itself at the perfect distance from the Sun
to support liquid water on its surface.
It's not too hot or too cold. It's just right.

If our planet were too close to the Sun, like Venus, warmer temperatures would eventually cause all of the water on the surface to evaporate. If our planet were too far from the Sun, like Mars, the colder temperatures would eventually cause all the water to freeze. Our planet seems to be located within a sort of temperate zone, which allows liquid water to persist and flow freely over our surface, which is a very good thing for us. Without liquid water, life as we know it cannot exist.

Actually, there is a combination of factors that allows our planet to support water—and we are talking about a lot of water! Over 70 percent of our planet is covered with it. Not only is the temperature warm enough in most places to allow water to remain liquid, our atmosphere is also dense enough to keep the water from evaporating in great quantities. While some water does evaporate, the planet's gravity is strong enough to keep it from escaping into space.

There is currently no other place in the solar system where liquid water can exist on the surface of a moon or a planet. There are hints that large salty oceans exist deep beneath the crusts of Europa, Ganymede, and Callisto, three of Jupiter's larger moons. Evidence for these oceans is only circumstantial, and because these oceans lie beneath miles of frozen crust, they will be extremely difficult, if not impossible, to explore. So for life as we know it, for the time being, you had better stick to Earth—it's where the water is.

What is the aurora borealis, popularly known as the northern lights?
Charged particles from the Sun interacting with
atoms in Earth's upper atmosphere.

Usually, charged particles of solar wind are deflected by Earth's magnetic field before they actually hit our planet. However, when there is a large influx of particles, such as after a solar flare or a coronal mass ejection, Earth's magnetic field is overwhelmed, so large quantities of charged particles leak through. These particles follow the magnetic field lines and are carried toward the North and South Poles of our planet. There, these particles interact with atoms in the upper atmosphere, causing the dancing, glowing lights known as the aurora borealis—the northern lights—and the aurora australis—the southern lights.

Normally, the northern lights are visible only in the upper latitudes (Alaska, Canada, northern Europe, and Russia). However, when there is a particularly large solar eruption, it is possible for the aurora to be seen as far south as the middle latitudes (California or Kansas, for example).

What is the ozone layer and what's wrong with it?
*A protective layer in Earth's upper atmosphere
that blocks ultraviolet radiation from the Sun.
Because of human influence, this layer is slowly decaying
and allowing harmful radiation through to the surface.*

Ozone is a molecule made up of three oxygen atoms (O_3). High in Earth's upper atmosphere, a layer of ozone molecules acts as a barrier that blocks high-energy ultraviolet radiation from the Sun and keeps it from reaching the surface.[2] Unfortunately, this natural layer of ozone is slowly eroding, and it appears that humans are responsible for its demise.

Over the past few decades, humans have been releasing various materials into the atmosphere, including chlorofluorocarbons (CFCs). CFCs have been used as coolants in refrigerators (Freon, for example) and propellants in aerosol cans. When released into the air, CFCs rise into the upper atmosphere where they break down into their individual atoms. The newly released chlorine atoms interact with the ozone, causing it to break down into oxygen (O_2) and chlorine monoxide (ClO). While this is a natural reaction, such large quantities of chlorine should not have access to the ozone layer. This extra chlorine is destroying ozone at a much faster rate than nature is capable of replenishing.[3]

The main damage to the ozone layer can be found above the continent of Antarctica, where a combination of wind patterns and cold temperatures cause CFCs to collect over the South Pole. Where large numbers of CFCs break down the ozone, a hole results. Over the past few years, the hole in the ozone layer has grown in size.[4] It is now large enough that, in addition to the entire continent of Antarctica, the southernmost tip of South America is starting to feel the effects of an increase in radiation.

Loss or even further erosion of the ozone layer will have serious ramifications for all life on Earth. For the human race, an increase in ultraviolet radiation will result in a dramatic rise in the number of skin cancers and eye problems. High enough levels of radiation can even result in death. Animals and plant life will be affected as well. In fact, the entire ecosystem will suffer, since life on this planet was not designed to live with such high radiation levels. Because destruction of the ozone layer is a global problem, it will require the united efforts of all the countries of the world to work toward a solution. Some headway has been made to regulate the use of these problematic products and destructive human activities, but as the hole continues to grow in size, stricter measures may be needed.

Why does the Sun move across the sky?

The Sun only "appears" to be moving across the sky, rising in the east and setting in the west. What is actually moving is Earth. Our planet is spinning around on its axis, while the Sun remains stationary.

Earth takes twenty-four hours, or one day, to rotate on its axis. It is this rotation that causes the Sun to "travel" across the sky. To help put things in perspective, think about the last time you rode in a car. As you traveled down the road, it looked as though things were moving outside your window while you were sitting perfectly still. As passengers on Earth, we are along for the same kind of ride through space. The only way we sense our motion is to see things "outside" the Earth—the Sun and stars—move. As Earth turns, they seem to move in and out of view.

A simple demonstration can help explain things further. Find a lamp in your room. (A ceiling light will work, but it is better to have a light at eye level.) For this demonstration, the light will be the Sun and you will be Earth. If you like, your nose can represent the house you live in.

Stand directly in front of the light, facing it. This represents noon. On Earth, the Sun would be straight overhead. Now, Earth (that's you) rotates on its axis. Start turning to your right. What happens to the Sun? It appears to move. Stop when your left shoulder is nearest the Sun. This represents sunset.

Keep turning. You can't see the Sun at all now. This is nighttime. Everything you see in the room now represents stars you could see at night. As you continue to spin on your axis, soon the Sun will show up over your right shoulder. This represents sunrise. Keep turning without stopping, from sunrise to sunset. Even though it looks as if the light is moving, Earth (yourself) is the one doing all of the work.

What causes the seasons?
Earth's tilted axis.

During summer in the Northern Hemisphere, it may seem that Earth is moving closer to the Sun as it gets hotter and hotter. This is not the case at all. In fact, during a Northern Hemisphere summer, Earth is actually slightly farther away from the Sun than it is during winter. It turns out that our planet's distance from the Sun doesn't directly affect the hot and cold of the seasons. Instead, Earth's seasons are the result of its tilted axis.

Earth's axis is not pointed straight up and down as it travels around the Sun. It's tipped over a little.

The tilted North Pole of our planet is always pointed toward a star named Polaris (the North Star). This doesn't change as Earth moves around the Sun. During summer in the Northern Hemisphere, the North Pole is tilted in the direction of the Sun as it points to Polaris. That extra tilt allows the Sun to rise higher in the northern skies and brings more sunlight to that hemisphere. Six months later, Earth has moved along its orbit so that the North Pole is tilted away from the Sun, though still pointed at Polaris. Less sunlight reaches the Northern Hemisphere, and the Sun itself remains rather low in the sky. Winter has come to the north.

Why is it cold in winter and hot in summer?
Because of Earth's tilted axis.

When Earth's North Pole tilts away from the Sun, the Northern Hemisphere doesn't get as much sunlight. So, it gets cold.

When Earth's North Pole is pointed in the direction of the Sun, the Northern Hemisphere warms up and summer arrives. During winter in the Northern Hemisphere, Earth is actually about 3 million miles (4.8 million kilometers) closer to the Sun than it is during the summer. Earth's distance from the Sun doesn't affect the seasons. It's the tilted axis that brings us snow in the winter.

If it's winter in North America, why is it summer in Australia?
Once again, because of Earth's tilted axis.

During winter in the Northern Hemisphere, the North Pole is tilted away from the Sun. If the North Pole is tilted away from the Sun, where is the South Pole? Pointed toward the Sun, of course. So the seasons are reversed. It is summer in the Southern Hemisphere, including Australia.

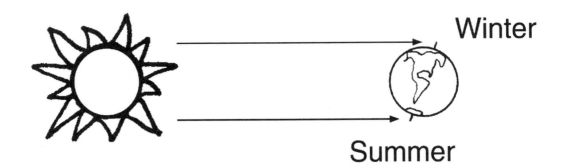

What is the winter solstice?
The name for a point in Earth's orbit when the North Pole is pointing as far away from the Sun as it ever gets.

There are many ways to describe the winter solstice. The winter solstice marks the first day of winter in the Northern Hemisphere. It is also the day that the Sun appears the farthest south in our skies. And, it marks the shortest day and longest night for the Northern Hemisphere.

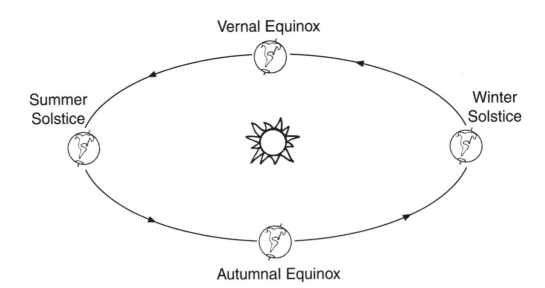

What is the summer solstice?

*The name for a point in Earth's orbit when the
North Pole is pointing as close to the Sun as it ever gets.*

As with the winter solstice, there are many ways to describe its counterpart, the summer solstice. The summer solstice marks the first day of summer in the Northern Hemisphere. On this day, the Sun appears the farthest north in our skies. It also marks the longest day and shortest night for those in the Northern Hemisphere.

What is the vernal equinox?

*The name for a point in Earth's orbit when the North Pole
is pointing exactly ninety degrees from the Sun as
Earth makes its way toward the summer solstice.*

The vernal equinox is a day of equal sunlight and darkness—when the Sun rises directly in the east and sets directly in the west. It also marks the first day of spring in the Northern Hemisphere. As Earth travels around the Sun, there are two points in its orbit where there is a day of equal light and dark. One day is the vernal equinox and the second day is the autumnal equinox.

What is the autumnal equinox?

*The name for a point in Earth's orbit when the North Pole
is pointing exactly ninety degrees from the Sun as
Earth makes its way toward the winter solstice.*

The autumnal equinox marks the first day of autumn in the Northern Hemisphere. This day is also marked by equal amounts of sunlight and darkness, as the Sun rises directly in the east and sets directly in the west.

ABOUT THE MOON

What is a Blue Moon?

What is the Moon made of?

How far away is the Moon?

How was the Moon formed?

What formed the craters on the Moon?

How old are rocks on the Moon's surface?

Why does the Moon change shape?

What are the different phases of the Moon?

What is a lunar eclipse?

When is the next lunar eclipse?

Can you always see the Moon?

What is the Man on the Moon?

Does the same side of the Moon always face Earth?

What is the "far side of the Moon"?

Is there a dark side of the Moon?

What is a Blue Moon?
The second Full Moon in a month or the third of four Full Moons in a season.

With such a strange name, one might expect an equally strange description. In the case of a Blue Moon, however, the name is really more exotic than the actual event. While it seems the term "Blue Moon" has been around for centuries (usually referring to something impossible or something that will rarely happen), it has only recently gained the commonly accepted definition of the second Full Moon within a calendar month—and that meaning appears to have come about by accident.[1]

In the nineteenth and early twentieth centuries, the *Maine Farmers' Almanac* described a Blue Moon as the third of four Full Moons in a season. (Actually, it's a little more complicated than that, having to do with Gregorian calendars, what day Lent falls on, and a few other strange twists.)[2] However, a mistake in a popular astronomy magazine in 1946 reported that a Full Moon was the second Full Moon of a month.[3] The media picked up on this definition and a "new" Blue Moon entered our culture.

According to this new description, a Blue Moon can occur because the Moon takes approximately twenty-nine and a half days to travel from one Full Moon to the next. Since all the months in the calendar (except February) have thirty or thirty-one days, it's just a matter of time until one Full Moon falls at the beginning of the month and the next one at the end of the month. As a result, a Blue Moon isn't as rare as the expression "once in a Blue Moon" might suggest. It turns out that a Blue Moon occurs about once every two or three years. The next three Blue Moons occur on June 30, 2007; December 31, 2009; and August 31, 2012.[4]

The Full Moon. Image credit: © UC Regents/Lick Observatory

Many Native American

tribes named each Full Moon according to the time of year it was visible. While they didn't have a Blue Moon, they did have many other types of moons, including Wolf Moon, Snow Moon, Worm Moon, Pink Moon, Flower Moon, Strawberry Moon, Buck Moon, Sturgeon Moon, Barley Moon, Harvest Moon, Hunter's Moon, Beaver Moon, and Cold Moon, to name a few.[5]

In modern times, many of these names have fallen out of use. However, the Harvest Moon and the Hunter's Moon are still used here and there. The Harvest Moon is always the Full Moon that occurs closest to the autumnal equinox—the first day of fall in the Northern Hemisphere. The Hunter's Moon is the Full Moon that follows the Harvest Moon.

What is the Moon made of?
Rocks.

Although Earth and the Moon are made up of similar elements, there are a few striking differences.

One major difference between the two is water—or lack of it. While water is plentiful on Earth, there isn't a drop of it on the lunar surface. Not only that, samples brought back by the Apollo astronauts indicate that there wasn't any water present during the formation of lunar rocks.[6] This fact suggests that there never has been any water on the Moon. While the Moon lacks water, it has an abundance of other things. There are higher levels of calcium, aluminum, titanium, uranium, and thorium on the lunar surface than can be found on Earth.[7]

**Exploring the Moon.
Image credit: NASA**

Most of the rocks on the Moon are igneous and basaltic in nature, meaning they were formed by cooling lava.[8] This lunar lava did not come from volcanoes, however, because there are no volcanoes on the Moon. Instead, it slowly seeped up to the surface through cracks in the Moon's crust. Other rocks, called *breccia* (breh'-cha), were formed during meteor impacts. The heat from the impacts caused smaller rocks to melt and stick together, forming new, larger rocks.

How far away is the Moon?
The Moon's average distance from Earth is 238,700 miles (384,400 kilometers).

The Moon's distance from Earth can vary by as much as 26,200 miles (42,200 kilometers). At its closest point, the Moon is 225,600 miles (363,300 kilometers) from Earth. At its farthest, the Moon lies at a distance of 251,800 miles (405,500 kilometers).[9] These distances are all measured from the center of Earth to the center of the Moon. Using the centers of each body allows scientists to be consistent with their measurements, without worrying about which mountain top they measured from or what crater floor they measured to.

How was the Moon formed?
When a Mars-sized object slammed into Earth, the resulting debris was trapped in orbit around our planet where it eventually coalesced into the Moon.

Scientists don't really know for sure exactly how the Moon formed since they can't travel back through time and watch. However, by comparing the compositions of both the Moon and Earth, they have pieced together what probably happened. Many scientists believe that the Moon was formed when an object about the size of Mars slammed into Earth soon after our planet formed. The tremendous impact flung huge amounts of debris into space, debris from both Earth and the impacting body. Some of this debris was blown completely away from our planet, but most remained trapped in Earth orbit forming a Saturn-like ring system. Over time, this debris—thousands of pieces of rocks and dust—gradually came together to form one solid object, our Moon.[10] (Saturn's rings have shepherd moons that keep them in place.)

When comparing the Moon and Earth today, the differences in the Moon's composition can be blamed on the debris from the large body that hit Earth. The similarities can be explained by the debris that came from Earth itself. If the Moon was originally a chunk of Earth that had somehow been ripped away from the planet, the two should have the same composition—but they don't. If the Moon had formed in some other place in the solar system and then became trapped by Earth's gravity, its composition should be much different from Earth's—but it's not. The impact theory seems to fit the Moon we have today.

What formed the craters on the Moon?
Explosions caused by meteor impacts.

Because the Moon doesn't have an atmosphere to protect it, every piece of space debris aimed in its direction crashes into the lunar surface. The resulting explosion flings debris in all directions. When the dust settles, the Moon has a new hole in the ground—a crater.

For years scientists were confused by craters on the Moon. They didn't understand why all of the craters were round. Think about this for a minute: if you hold a rock at arm's length and let it drop, it will make a fairly round crater in the ground. However, if you throw the rock and it hits the ground at an angle, the resulting crater will be shaped more like an oval. The greater a rock's angle of impact, the more oval the resulting crater. At some point, the rock may even skip across the ground a couple of times before coming to a rest. On the Moon, there were no signs of oval craters or skip marks. Since it seemed highly unlikely that all the meteors struck the Moon straight on, there had to be something else going on during an impact, something that would always leave behind a round crater.

Rock samples brought back from the Moon showed all the signs of having been formed by large impacts. Scientists began conducting a variety of impact experiments in an effort to explain what they were seeing. They soon realized that when a large, fast-moving rock impacts a solid surface, there is a tremendous amount of energy released. The explosion that results would fling equal amounts of material in all directions, leaving a round crater at the point of impact. In other words, it doesn't matter at what angle a meteor hits, the resulting explosion will almost always leave a round crater.

Close-up images of the Moon taken by orbiting spacecraft eventually revealed some small oval craters and even some skip marks—all of these too small to be seen by Earth-based telescopes. Scientists believe these oval craters are actually secondary craters, craters caused by material that was flung up by a meteor impact and then fell back to the lunar surface. When this debris hit the ground, it wasn't traveling nearly as fast as the original meteor, therefore there wasn't nearly as much energy involved and no explosion.

How old are rocks on the Moon's surface?
Three to four and a half billion years old.

Apollo astronauts brought back many samples of rock from various parts of the Moon. The oldest samples were found in the lighter-colored, heavily cratered regions. One sample turned out to be 4.44 billion years old. The youngest rocks were a mere 3.1 billion years old. These youngsters were found in the lunar maria, the smooth, dark areas on the lunar surface.[11]

The ages of the lunar rocks suggest that

Collecting samples of the lunar surface.
Image credit: NASA

most of the large features we see on the Moon formed early in its life. While there continue to be impacts on the lunar surface, they are much smaller, and the resulting craters are rarely large enough to be seen from Earth. In other words, the Moon's appearance has changed little in the last three billion years.

Why does the Moon change shape?
It doesn't, but it seems to change because we see the Sun shining on different parts of the Moon as it travels around Earth.

Some people believe the different shapes, or phases, the Moon goes through are caused by Earth's shadow falling on different parts of the Moon at different times. This isn't the case. Instead, we are seeing different parts of the sunlit side of the Moon as it travels around Earth.

During a Full Moon (position #1), it looks as though you are seeing the entire Moon. In fact, you are seeing all of the Moon's sunlit side. You can see a Full Moon only when the Moon

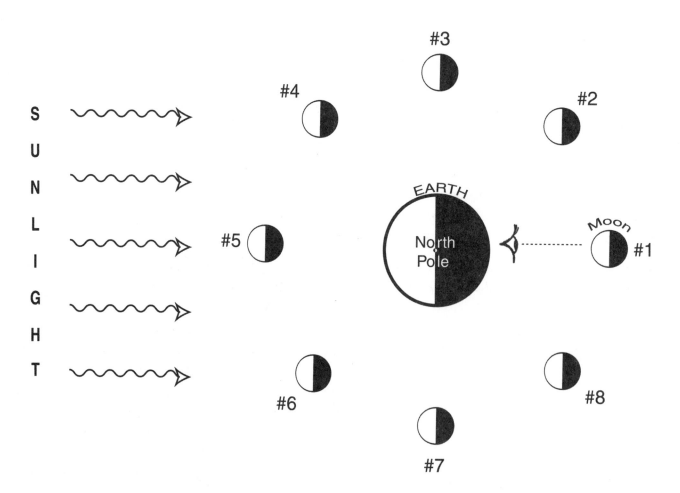

The Moon and its phases as seen from space—high above Earth's North Pole.

The Moon's phases as seen from Earth.
(The numbers below each phase correspond to numbers in the previous illustration.)

has reached the point in its orbit where Earth lies between it and the Sun. (It is during a Full Moon that lunar eclipses can occur.)

As the Moon travels around Earth, the Full Moon fades to a Waning Gibbous (position #2), then to a Third Quarter Moon (position #3). During a Third Quarter Moon, what you are really seeing is half of the sunlit side of the Moon and half of the side the Sun is not shining on.

After a Third Quarter Moon comes a Waning Crescent Moon (position #4). Then, as the Moon's orbit takes it between Earth and the Sun, we see less and less of the sunlit side until it completely disappears from sight. When we can't see any of the sunlit side of the Moon, this is what is called the New Moon (position #5). (It is during a New Moon that solar eclipses can occur.)

After the New Moon, the Moon continues to travel around Earth and the phases start over as we begin to see more of its sunlit side. First comes the Waxing Crescent phase (position #6), then First Quarter (position #7). Notice that while we are seeing half the sunlit side and half the dark side of the Moon during both First Quarter and Third Quarter, the light side and dark side are reversed.

After First Quarter comes a Waxing Gibbous Moon (position #8) and then another Full Moon. The cycle from Full Moon to Full Moon takes about twenty-nine and a half days.

What are the different phases of the Moon?
New Moon, Waxing Crescent, First Quarter, Waxing Gibbous, Full, Waning Gibbous, Third Quarter, Waning Crescent, and back to New.

A New Moon is invisible from Earth, because all of the Moon's sunlit side is facing away from our planet. When the Moon grows in size, the term "waxing" is used. "Waning" is used to describe a Moon that is getting smaller. "Crescent" describes a banana-shaped Moon, and "gibbous" is used when you can see more than half but not quite all of the Moon's disk.

What is a lunar eclipse?

A lunar eclipse occurs when Earth's shadow actually does fall on the Moon.

During a Full Moon, when Earth is between the Sun and the Moon, sometimes the Moon passes through Earth's shadow. When this happens, a lunar eclipse occurs.

A lunar eclipse occurs when a Full Moon passes through Earth's shadow.

If only a part of the Moon passes through Earth's shadow, it creates a partial lunar eclipse. If all of the Moon passes through Earth's shadow, it is a total lunar eclipse. Lunar eclipses occur only during a Full Moon, but they don't occur during every Full Moon. Most of the time, the Moon's orbit takes it either above or below Earth's shadow.

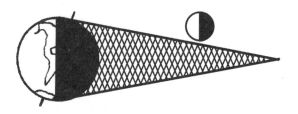

Most of the time, the Full Moon passes either above or below Earth's shadow.

When is the next lunar eclipse?

Listed in the chart below are the dates, times, and best viewing locations for all the partial and total lunar eclipses through the year 2010.

Lunar Eclipses for 2005–2010

Date	Time of Maximum Coverage*	Eclipse Type	Eclipse Duration†	Where Visible
2005 Oct 17	12:03 UT (8:03 AM EDT)	Partial	00h58m	Asia, Australia, Pacific, North America
2006 Sep 07	18:51 UT	Partial	01h33m	Europe, Africa, Asia, Australia
2007 Mar 03	23:21 UT (6:21 PM EST)	Total	03h42m, **01h14m**	Americas, Europe, Africa, Asia
2007 Aug 28	10:37 UT (6:37 AM EDT)	Total	03h33m, **01h31m**	Eastern Asia, Australia, Pacific, Americas
2008 Feb 21 (2008 Feb 20)	03:26 UT (10:26 PM EST)	Total	03h26m, **00h51m**	central Pacific, Americas, Europe, Africa
2008 Aug 16	21:10 UT	Partial	03h09m	South America, Europe, Africa, Asia, Australia
2009 Dec 31	19:23 UT	Partial	01h02m	Europe, Africa, Asia, Australia
2010 Jun 26	11:38 UT (7:38 AM EDT)	Partial	02h44m	Eastern Asia, Australia, Pacific, western Americas
2010 Dec 21	08:17 UT (3:17 AM EST)	Total	03h29m, **01h13m**	Eastern Asia, Australia, Pacific, Americas, Europe

*Times given in Universal Time (UT) and Eastern Standard Time (EST) or Eastern Daylight Time (EDT) when appropriate.

†First time indicates duration of eclipse from beginning to end. For total eclipses, second time (in bold letters) indicates time of total coverage.

The author would like to thank Fred Espenak, NASA Goddard Space Flight Center, for the use of his eclipse charts.[12] (For a list of lunar eclipses through the year 2020, see page 206.)

Can you always see the Moon?

No.

Sometimes you can see the Moon during the evening and sometimes you can see it during the day. However, there is a period during every month when you can't see the Moon at all. During the New Moon phase, when the Moon lies between Earth and the Sun, no part of the Moon's sunlit side is visible from Earth.

What is the Man on the Moon?
An imaginary figure composed of light and dark areas on a Full Moon.

Besides the Man on the Moon, stories from all over the world tell of different creatures found on the Moon. Some speak of a dragon, others see a rabbit, and one story tells of a man who sells cabbages! Without a telescope, the smooth dark areas (maria) and the lighter-colored, heavily cratered regions on the Moon blur together. With your imagination, what do you see when you stare at the Moon?

Does the same side of the Moon always face Earth?
Yes.

Whenever you see the Moon in the sky, you always see the same craters and smooth dark areas. This is because the same side of the Moon always faces our planet.

While we can see only 50 percent of the lunar surface at any given time, the Moon does wobble slightly in its orbit. These wobbles, called librations, allow us to see a total of 59 percent of the lunar surface over time.[13] These librations occur for a variety of reasons, including gravitational effects of the Moon's orbit around Earth and the uneven distribution of mass within the Moon.

What is the "far side of the Moon"?
The side of the Moon that never faces our planet.

From our vantage point on Earth, we can't see the far side of our nearest companion. The only way we can see the lunar far side is to send spacecraft around the Moon and have them take pictures. The first images of the Moon's far side were taken by the Russian *Luna 3* spacecraft on October 7, 1959.[14] These images and others show a landscape much different from the one we see every month. On the far side of the Moon, there are no maria (smooth dark areas) that cover large portions of its Earth-facing side. There are only craters, craters, and more craters.

Taken by *Apollo 16* astronauts, this image shows a large portion of the Moon's heavily cratered far side. The dark markings on the left are maria that can be seen from Earth.
Image credit: NASA

Is there a dark side of the Moon?
Yes and no.

At any given moment, half the Moon is light and half is dark. But as the Moon moves around Earth, different portions of its surface go from day to night and vice versa. No portion of the Moon is dark all the time.

The drawings to the right show the Moon in eight different parts of its orbit around Earth. In the top drawing, an "x" marks a crater that can be seen from Earth. An "o" marks a crater on the far side. Notice as the Moon travels around Earth, the "x" always faces Earth and the "o" always faces away from Earth. (From the previous questions, we know that the same side of the Moon always faces Earth.)

The next drawing adds shadows to the picture. We can't see any part of the Moon that the Sun is not shining on. If the "x" is in the sunlight, we can see the crater. If the "x" is in the dark, we can't.

As the Moon travels around its orbit, sometimes the "x" is in the sunshine and sometimes the Sun is shining on the "o." So you can see every that side of the Moon receives sunlight, just at different times.

ABOUT SPACECRAFT AND SPACE TRAVEL

How many space shuttles are there?
Four.

Currently there are three space shuttles that have active flight status: *Discovery*, *Atlantis*, and *Endeavor*. *Enterprise*, the first shuttle, is in the Smithsonian's McDonnell Space Hangar near Dulles Airport in Washington, DC.[1] *Enterprise* never flew in space. Instead, it was dropped several times from the back of a 747 airplane to test the shuttle's aerodynamic properties and to show astronauts how the spacecraft would handle after it reentered Earth's atmosphere.

An image of the space shuttle Endeavor *can be seen in the color insert section of this book.*

Originally, there were five space shuttles. As a test vehicle, *Enterprise* came first. However, *Columbia* was the first to launch into space. After five flights, *Columbia* was followed by *Challenger*, then *Discovery*, and finally *Atlantis*. The shuttle *Endeavor* was constructed to replace *Challenger*, which was destroyed shortly after liftoff on January 28, 1986. It is not likely that another shuttle will be built to replace *Columbia*, which broke apart during reentry on February 1, 2003.

Can the space shuttle travel to the Moon?
No.

The space shuttle was not designed to go to the Moon. Its main engines and solid rocket boosters provide only enough thrust to carry the shuttle into low Earth orbit. Once in flight, it circles above our planet at an average distance of 250 miles (405 kilometers).[2] Recall that the Moon's average distance places it over 238,700 miles (384,400 kilometers) from our planet.

Before the shuttle could travel to the Moon, engineers would have to figure out a way for it to leave Earth's orbit. Normally, the shuttle travels around Earth at speeds ranging from 17,000 to 18,000 miles per hour (27,500 to 29,100 kilometers per hour).[3] To escape Earth's orbit, it would have to accelerate to at least 25,039 miles per hour (40,320 kilometers per hour), a speed known as Earth's escape velocity.[4] The main engines of the shuttle are simply not powerful enough to produce that extra 8,000 miles per hour (12,820 kilometers per hour) for a body as massive as the shuttle.

Finding enough thrust is only the beginning of the problems one would encounter if trying to adapt the shuttle for Moon flight. The obstacles and costs involved would be extremely difficult, if not impossible, to overcome. It would be much better to start from scratch or examine the one rocket that did manage to take humans to the Moon, the *Saturn V*.

Launch of space shuttle *Discovery* on August 10, 2001.
Image credit: NASA

Does the space shuttle have to dodge stars?
No.

As stated previously, the space shuttle was designed to travel in low Earth orbit. Its average orbit takes the shuttle approximately 250 miles (405 kilometers) above the surface of our planet.

Now consider that the closest star to Earth—the Sun—lies 93 million miles (149 million kilometers) away from us. Beyond that, the next closest star system (Alpha Centauri) is about 25 trillion miles (40 trillion kilometers) away. All other stars are much more distant than that. When you compare the maximum distance the space shuttle can travel (600 miles) to the distance to Alpha Centauri (25 trillion miles), it becomes obvious that the shuttle doesn't have to worry about running into any stars!

What happened to the space shuttle *Columbia*?
On February 1, 2003, Columbia broke apart while passing through Earth's atmosphere on its way toward a landing at Kennedy Space Center in Florida. All seven crew members onboard were lost.

When a space shuttle reenters Earth's atmosphere, it experiences a tremendous amount of heat because of friction that builds up between the fast-moving shuttle and the atmosphere. During this critical time, temperatures along the shuttle's underside can soar as high as 3,000°F (1,650°C).[5] The spacecraft is protected from this searing heat by a layer of special heat-resistant tiles. It was damage to these tiles that caused *Columbia*'s destruction.

During *Columbia*'s launch on January 16, 2003, a suitcase-sized chunk of insulation broke off the shuttle's external tank and slammed into the leading edge of the shuttle's left wing. Normally, this insulation is as soft as a sponge. However, because the piece was traveling between 416 and 573 miles per hour (670 and 923 kilometers per hour) when it struck the wing, the insulation severely damaged the protective tiles in that area.[6] Soon after *Columbia* reached orbit, it is believed that pieces of the damaged tile (or tiles) broke away, leaving a hole in the shuttle's wing. This damage did not affect the shuttle during its sixteen-day mission. However, the moment *Columbia* began to reenter the atmosphere, temperatures along its leading edges began to rise and the damaged area began allowing superheated atmospheric gases to enter the shuttle's wing.

These hot gases overheated metal supports within the wing and left wheel well. Soon the left wing, weakened by the intense heat, broke apart or collapsed, causing the shuttle to veer sharply off its flight path. Traveling at supersonic speeds, the shuttle was unable to withstand the resulting forces and broke apart.[7] Unfortunately, all seven crew members were lost. Amazingly, no one on the ground was hurt by the debris that rained down across several southern states.

Columbia's final crew consisted of the following individuals: Rick D. Husband, commander; William C. McCool, pilot; Michael P. Anderson, payload commander; David M. Brown, mission specialist; Kalpana Chawla, mission specialist; Laurel Blair Salton Clark, mission specialist; and Ilan Ramon, payload specialist.[8]

What happened to the space shuttle *Challenger?*

On January 28, 1986, Challenger *broke apart shortly after launch when a rubber O-ring within a solid rocket booster failed, allowing hot gases to destroy the external tank. All seven crew members were lost in the resulting explosion.*

After several launch delays because of weather and technical difficulties, *Challenger's* last mission began on a cold January morning. Just seventy-three seconds into the flight, the spacecraft appeared to explode.[9] The solid rocket boosters eerily continued on, flying off in two different directions before they were destroyed by ground controllers. A review board determined that it was a failure within one of these rocket boosters that resulted in *Challenger's* destruction.

Each shuttle has two large solid rocket boosters attached to its large external tank. In addition to the liquid fuel main engines, the boosters ignite at launch and provide the extra thrust needed to lift the shuttle into orbit. Because the solid rockets are so large, each booster is made up of several sections. Within the joints that connect the different sections, there are circular rubber gaskets called O-rings that seal the different parts together. It was one of these O-rings that proved to be sensitive to cold weather.

The morning of the *Challenger* launch, the temperature at the Kennedy Space Center hovered just around freezing. When the boosters ignited, one of the cold O-rings was unable to expand properly. Hot gases forced their way through the cold gasket and superheated the external tank, causing it to fail. As a result, over two million pounds of hydrogen was suddenly released into the atmosphere in an explosive rush, forcing *Challenger* off its flight path. The shuttle, traveling at supersonic speeds, was broken apart by aerodynamic forces.[10]

Challenger's final crew consisted of the following individuals: Francis R. Scobee, commander; Michael J. Smith, pilot; Judith A. Resnik, mission specialist; Ronald E. McNair, mission specialist; Ellison S. Onizuka, mission specialist; Gregory B. Jarvis, payload specialist; and S. Christa McAuliffe, the first teacher in space.[11]

Why do we need the International Space Station?
If the human race is going to explore outer space, we need to know much more about it.

Space, with its microgravity and no air to breathe, is a very different environment from what we are used to on Earth. The International Space Station, as well as the previous Skylab, Soyuz, and Mir space stations, has greatly improved our knowledge of living and working in space, but there is so much more we need to comprehend about this unique environment. Also, while trying to understand the environment of outer space, we gain a better appreciation of the environment we live in on Earth.

An image of astronauts working on the International Space Station can be found in the color insert section of this book.

The space shuttle allows astronauts to conduct many experiments in the weightlessness of space. But because the shuttle missions are limited to a span of just one or two weeks, the experiments must be kept short. The International Space Station, once completed, will give scientists the time needed to conduct long-term experiments. In addition, they can study the long-term effects of weightlessness on the human body and the reactions and behaviors of crew members in such cramped quarters. All the information gleaned from the International Space Station will be useful in other missions, whether we are developing a colony under the sea, planning a colony on our Moon, or traveling to another moon or planet.

How many astronauts have walked on the Moon?
Twelve.

For each of the Apollo missions, there were three astronauts. Two landed on the lunar surface while one remained onboard the command module in lunar orbit.

Apollo 11, 12, 14, 15, 16, and *17* landed on the Moon. An explosion onboard *Apollo 13* while on its way to the Moon made a lunar landing impossible. *Apollo 13* astronauts only flew by the Moon once on their return to Earth.

The astronauts who walked on the moon were *Apollo 11*'s Neil Armstrong and Edwin "Buzz" Aldrin, *Apollo 12*'s Charles (Pete) Conrad and Alan Bean, *Apollo 14*'s Alan Shepard and Edgar Mitchell, *Apollo 15*'s David Scott and Jim Irwin, *Apollo 16*'s John Young and Charles Duke, and *Apollo 17*'s Eugene Cernan and Harrison (Jack) Schmitt.[12] (For more information about the manned and unmanned missions to the Moon, see page 227.)

From Earth, can we see the astronauts' footprints on the Moon?
No, not even through the most powerful telescope.

Using the largest telescopes on Earth, the smallest object we can see on the Moon is about half a mile (one kilometer) across. The Apollo astronauts wore big boots, but not that big!

Could we travel to the Moon tomorrow?
No.

True, we have been to the Moon, but the last mission took place in 1972, over thirty years ago. The only rocket powerful enough to take humans to the Moon was the *Saturn V*. Today, there are only three of those rockets left and they are on display as museum pieces at the Kennedy Space Center in Florida, Johnson Space Center in Texas, and the Marshall Space Flight Center in Alabama.

In January 2004 President Bush declared that the US space program would once again head to the Moon. Even with

**Footprint on the Moon.
Image credit: NASA**

money budgeted for this program, it is going to take a while to get there. Why? Because engineers and scientists are having to start nearly from scratch.

While we certainly have the technology to go to the Moon, that technology is forty years old. Would you want to use a computer that was forty years old? It will take time to redesign everything to take advantage of today's technology.

Why should we go back to the Moon?
The Moon would provide an excellent training ground for future planetary missions. In addition, the lunar soil contains elements that might provide opportunities for mining. The Moon could also be an exotic location for adventure-seeking vacationers. The possibilities are endless.

If Apollo astronauts had discovered vast quantities of precious metals on the lunar surface, the resulting gold rush might have resulted in mining colonies and other industries springing up all over the Moon. However, no such discoveries were made, and many people have adopted the attitude "been there, done that" when it comes to lunar exploration.

Yes, the human race has been to the Moon, but think about the numbers for a moment: twelve men spent a total of 12.5 days (less than two weeks) on the surface, exploring six landing sites and bringing back 836 pounds (379 kilograms) of samples.[13] In other words, there's still a lot more Moon out there to investigate.

In addition, if humans are planning to explore the other planets and moons of our solar system, it makes sense to practice establishing colonies on a world that is only three days away from home. By comparison, Mars is—at the very least—six months away.

Excavation of the lunar soil during the building of these colonies would provide geologists with a better understanding of the Moon's mineral content and what, if any, mining operations might be possible. Then, once these colonies were established, you can bet that there would be people willing to pay for a chance to visit.

Yes, there are definitely obstacles to overcome, but the Apollo program showed that it *can* be done. The problem today is finding the money to make it happen. Currently, the government provides the necessary funding for our space program. And while it does support space exploration, so much more money will be necessary to undertake more challenging missions. The funds required for a lunar colony will probably have to come from private industry, and that won't happen until someone figures out a way to turn the Moon into a moneymaking opportunity.

If we build colonies on the Moon, could we see them from Earth?
Perhaps, but only with a telescope.

Whether or not we could see these colonies depends on how large the colony is and if it is constructed of materials that reflect sunlight.

The first colonies would be very small. Remember, everything has to be brought from Earth. In addition, most of the colonies would be underground to protect the inhabitants from meteor impacts and radiation from the Sun. What little is left above the surface would probably be mining equipment or extra supplies. None of these things would be big enough to be seen from Earth, with or without a telescope.

Perhaps in the distant future, cities on the Moon will be large enough to be seen from Earth with a telescope. For something human-made to be seen from our planet without a telescope, it would have to be extremely large—at least 200 miles (320 kilometers) across.[14] Even Los Angeles isn't that big! To give you an idea about just how small an object that size would appear from Earth, go outside and look up at the Moon. Two hundred miles is about the size of the smallest lunar maria (dark area) you will be able to see without the use of binoculars or a telescope.

What was the first spacecraft to land on another celestial body?
Luna 2.

Luna 2, launched by the Soviet Union, crash-landed on the Moon's surface on September 14, 1959.[15] It was the first spacecraft from Earth to land (however hard) on another body of the solar system. Even though it crashed into the Moon, it successfully completed its mission—which was simply to make it to the Moon. Several missions before *Luna 2* had not even made it out of Earth's atmosphere. Almost five and a half years after *Luna 2*'s mission, *Luna 9* became the first spacecraft to do a controlled landing on a celestial body other than Earth.[16] *Luna 9*'s target was also the Moon.

What spacecraft carried animals to the Moon and back?
Zond 5

In an effort to study what effects (if any) traveling to the Moon and back would have on living beings, Soviet scientists placed turtles, worms, flies, and other small creatures in a small, protected payload compartment of the *Zond 5* spacecraft.[17] Launched on September 15, 1968, the spacecraft flew around the Moon three days later and headed back to Earth. It splashed down in the Indian Ocean on September 21, 1968.[18] The turtles had lost a bit of weight, but otherwise seemed fine. Not much is known about the condition of the other creatures, although it is doubtful they survived the forces of reentry.

Zond 5 was believed to be one of the final test missions before the Soviet Union planned to send a manned mission to the Moon. However, six of the next eight Soviet lunar missions (all robotic) exploded just after liftoff. These and other problems essentially ended the country's attempts to be the first to put humans on lunar soil. After *Apollo 11*'s Neil Armstrong and Buzz Aldrin landed on the Moon on July 20, 1969, the Soviets continued to send robots to the Moon but at a much slower pace. The last Soviet lunar mission, *Luna 24*, launched on August 9, 1976, and returned to Earth several days later with samples of lunar soil.[19]

What was the first spacecraft to land on another planet?
Venera 7

Just two months before Neil Armstrong became the first human to walk on the Moon, the Soviet Union's *Venera 7* became the first robotic spacecraft to land on another planet. *Venera 7* landed on Venus on May 17, 1969, and managed to transmit twenty-three minutes' worth of data before succumbing to the tremendous heat and pressure found on the Venusian surface.[20] Despite hellish conditions—an average temperature of 900°F (480°C) and pressures ninety times greater than Earth's pressure at sea level—the Soviet Union had eight successful *Venera* landers that collected a total of nine and a half hours' worth of data from the surface of Venus.[21]

(For more information about missions to the Moon, Venus, and the other planets of our solar system, see "A Brief History of Lunar and Planetary Exploration" beginning on page 225.)

Why haven't humans traveled to other planets?
Because it would take a really long time to get there.

It would take a long time, and a great deal of fuel, food, and air to breathe, in addition to an extremely large spacecraft in which to carry everything. However, the most important item required for a trip of this magnitude is money. Such a trip would probably be too expensive for any nation to undertake alone either now or in the foreseeable future.

The major problem with interplanetary travel, besides the cost, is the vast distances involved. It took Apollo astronauts three solid days of travel to get to the Moon. While that may not seem like much, think of it in terms of traveling in a car. Could you travel for three days and nights in a car without stopping for food or gasoline, or to stretch your legs? Remember, there are no restaurants between here and the Moon!

It takes three days to get to the Moon and three days to get back. If you want to spend time on its surface, it takes even longer, but humans have managed to accomplish this feat. If you wanted to go to another planet, like Mars, it would take much longer. A trip to Mars would take over six months. Once there, you would have to wait several additional months until Earth and Mars were properly aligned again before you could return home. (Because Earth travels around the Sun almost twice as fast as Mars, the two planets come close to each other only once every two years or so.) In total, a round trip to Mars could take almost two years—and Mars is one of the closest planets. It would take years to reach Jupiter, Saturn, and the other outer planets.

Should humans travel to Mars?
Yes.

It is human nature to explore —and of all the other planets in our solar system, Mars has the greatest potential to support human life. Even though astronauts would have to deal with below-freezing temperatures, no air to breathe, and almost no atmosphere, engineers already have plans for traveling to the Red Planet.

Several ideas have been suggested on how a Mars mission could be accomplished. One plan is to send—in advance—the food and supplies needed for the stay on the planet and the return trip. That way not everything will have to be crammed aboard the crew ship.

Once on the surface, scientists think that astronauts will be able to use the Martian soil for some of their building sup-

**Mars—as seen by the Hubble Space Telescope.
Image credit: Hubble Heritage Team (NASA/AURA/STScI),
James Bell (Cornell Univ.), Michael Wolff (Space Science Inst.)**

plies. Based on information sent back from the robotic rovers and landers that were already sent to Mars, scientists believe that astronauts will be able to make plaster of paris, ceramics, cement, glass, and maybe even blasting explosives out of the Martian soil. All these materials could be used in the construction of buildings, which would cut down on the amount of material that would have to be brought from Earth.

Again, a key factor in deciding if we travel to Mars is money. Scientifically and technologically, we can meet the challenge. Financially, the challenge will be a bit tougher to overcome.

Can we travel to the stars?
No, not yet.

The stars that you see at night are very, very far away from us. As noted earlier, Alpha Centauri is the closest of those stars and it is 24.6 trillion miles (39.7 trillion kilometers) away. If you got in your car and somehow managed to drive to Alpha Centauri following a sixty-five miles per hour speed limit, it would take you forty-four million years to get there.

If, however, you were traveling as fast as the *Voyager 2* spacecraft—which is flying out of our solar system at a speed of about 38,500 miles per hour—it would still take you almost seventy-three thousand years to get to Alpha Centauri.[22] Remember, that is just the closest star! We still haven't been able to send humans to the closest planet.

Do any of the spacecraft we send to the planets ever return to Earth?
No.

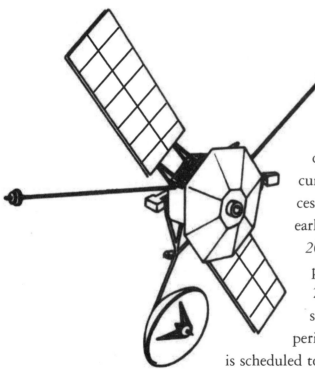

Very few of the spacecraft that are launched to other worlds will ever return to Earth. The few exceptions are special "sample return" missions, and to date none of these missions has ever been sent to another planet.

As of early 2005, there have only been four successful robotic sample return missions, with two currently underway. The Soviets conducted three successful sample return missions to the Moon in the early 1970s. These missions (*Luna 16* in 1970, *Luna 20* in 1972, and *Luna 24* in 1976) returned soil samples from the lunar surface.[23] On September 8, 2004, NASA's *Genesis* spacecraft returned samples of solar wind that it had collected over a two-year period.[24] In January 2006, NASA's *Stardust* spacecraft is scheduled to return samples it collected during its flyby of the comet Wild 2 on January 2, 2004.[25] In June 2007, the Japanese spacecraft *Hayabusa* is scheduled to return samples of the asteroid Itokawa it will collect during a June 2005 encounter.

Most of the time a spacecraft is designed to do all of its work "on location," so to speak, whether it is taking images as it flies by a planet or recording spectroscopic measurements of rocks as it roves around the Martian surface. These spacecraft record the data onto onboard computers and then relay them back to Earth via radio signals. Once their mission is complete and all of their data has been downloaded to Earth, there is no need for the spacecraft to return to its planet of origin.

While it might be nice to have an old *Lunar Orbiter* or the *Spirit* rover for a museum piece, cost is a major factor in the decision not to have a spacecraft return to Earth. A spacecraft designed to come back home would be heavier at launch because of the extra fuel and instruments needed for its return flight. The heavier the object you want to launch, the more powerful your rocket must be. As one might imagine, the more powerful a rocket is, the more it costs. In addition, there would be the expense of maintaining a support crew to monitor and control the spacecraft during its return flight.

The engineering challenges of a planetary sample return mission are daunting. The spacecraft must first successfully land on the planet, then collect and preserve its sample, launch its return package from the planet's surface, and be guided back to Earth. If all of that is accomplished, it still has to survive the fiery passage through our atmosphere and the landing. For the *Genesis* sample return mission, only the sample return package was sent back through the atmosphere for waiting scientists. (Even though the returning probe's parachute failed to open after it entered the atmosphere, scientists were able to retrieve many of the samples for study.) The primary spacecraft, after it had released its precious cargo, was diverted back into space and into a permanent solar orbit.

There have been plans to send a sample return mission to Mars for decades. However, costs and engineering hurdles have kept that project on the ground to date. Hopefully that mission will get underway soon so we can change the answer to the original question to yes.

How do we get all those beautiful pictures of the planets?
The same way you receive music through your radio
or a picture on your cellular phone—by using radio waves.

Engineers control a spacecraft by using radio signals. As with a wireless Internet connection, they transmit messages to and receive messages from the spacecraft's onboard computer using radio waves. (Remember, these are robotic spaceships with no humans onboard.) Controllers on the ground tell the spacecraft's computer where to aim cameras and other scientific instruments. They provide information about exposure times and how to move the spacecraft to avoid blurring long-exposure images. Because it takes time for the radio signals to travel through space, ground controllers must plan ahead and send the signals well in advance of the time that they want the spacecraft to execute their commands.

The cameras onboard the spacecraft are very similar to the digital cameras you can buy in a store. No film is used in the process. The image data is completely electronic and sent over the radio waves back to Earth once an exposure is finished. On Earth, the signal is received by the Deep Space Network. This network consists of three large radio antennas that are located roughly 120 degrees apart. Positioned around the globe, these antennas allow the controllers to be in constant communication with the spacecraft. (As Earth rotates one antenna out of the spacecraft's range, another antenna is rotating into view.) These antennas are located in Goldstone, California; Madrid, Spain; and Canberra, Australia.[26]

Once the electronic data are received by the Deep Space Network, computers on Earth translate the information into images. These images are then released to the public for all to see.

ABOUT STARS

Is it true stars twinkle and planets don't?

What is the North Star?

Will the North Star always show us the way north?

Is the North Star the brightest star in the sky?

Is there a South Star?

What is the Morning Star?

What is the Evening Star?

How do they name stars?

Do stars really fall?

Why do stars come out only at night?

How many stars can you see with your eyes alone?

Do people in different parts of the world see the same stars at the same time?

Do stars come in different colors?

What is a constellation?

How many constellations are there?

Do astronomers study constellations?

Can you see a constellation through a telescope?

How can I find a constellation?

What are the constellations of the zodiac?

How do astronomers measure the distances to stars?

What is a light-year?

What is a parsec?

What is the closest star to Earth?

How does a star form?

Will a star live forever?

How long do stars live?
What is a binary star?
What is an open cluster?
What is a galactic cluster?
What is a globular cluster?
What is a brown dwarf?
What is a red giant?
What is a red supergiant?
What is a white dwarf?
What is a nova?
What is a supernova?
What is a neutron star?
What is a pulsar?
What is a quasar?

Is it true stars twinkle and planets don't?
No and yes.

The saying that stars twinkle and planets don't is not correct. Neither stars nor planets actually "twinkle." However, we notice apparent changes in stars more than we do in planets. So the second answer is yes; most of the time stars appear to twinkle while the planets don't.

The twinkling effect is not caused by the star or planet. Instead, it is caused by Earth, or more specifically, Earth's atmosphere. Light from a star or planet doesn't begin to twinkle until it passes through the turbulent air that we breathe. On its way down to the ground, it can get jostled by wind, water vapor, dust, and other contaminants in the air. Our eyes detect this jostling (or twinkling) more in the light from distant, pinpoint stars than we do in the brighter light from the disks of nearby planets. In other words, planets "twinkle" just as much as stars, but because they are brighter and closer to us, we don't notice the movement as much. If Earth didn't have an atmosphere, all of the stars and planets would appear as constant steady points of light.

What is the North Star?
A star called Polaris that is currently lined up with Earth's North Pole.

Have you ever seen someone spin a basketball on his finger? As the basketball spins around, the middle of the basketball spins around in a big circle. The area under his finger is spinning, too, but in very tiny circles.

Now, think of Earth as the basketball. Instead of spinning around on a finger, our planet spins around on its axis. Earth's axis is represented by the North and South Poles. The person's finger would represent the South Pole and is pointing up through the North Pole and on to the star Polaris. As Earth turns, the North Pole continues to point to Polaris while Earth's surface rotates away from all of the other stars.

Will the North Star always show us the way north?
No.

Polaris, our North Star, will guide us north for the rest of our lifetimes and that of our great-great-grandchildren as well. So for us, Polaris could be considered a permanent North Star. But, in a few thousand years, that will change.

Polaris is the North Star because Earth's North Pole points in its direction. However, the direction in which Earth's North Pole points changes over time. This movement, called precession, slowly moves the North Pole in a huge circle, taking twenty-six thousand years to circle around once.[1]

Five thousand years ago the North Pole pointed to the star Thuban in the constellation Draco. In twelve thousand years, the North Pole will point near the star Vega in the constellation Lyra. In twenty-six thousand years, the North Pole will once again point to Polaris.[2]

Is the North Star the brightest star in the sky?
No, not by a long shot!

Since just about everyone has heard about the North Star, many people think it must be the brightest, most impressive star in the sky. That's not true. In fact, Polaris is just an average star. What makes it special is its location in the sky. If it were not lined up with Earth's North Pole, Polaris would simply be known as the brightest star in the faint constellation of Ursa Minor, the Little Bear.

The brightest star in the night sky is called Sirius. It is located in the constellation Canis Major, the Big Dog, and is sometimes called the Dog Star. The brightest star in the daytime sky is, of course, our Sun.

(For a list of the twenty-five brightest stars in the sky, see page 213.)

Is there a South Star?
No.

There is currently no star that lines up with Earth's South Pole, so there is not a South Star. The fact that there is a star lined up with the North Pole is coincidence.

What is the Morning Star?
Venus

Though not a star at all, the planet Venus appears as such a bright starlike point in the early morning skies, it has been mistaken for a "Morning Star" for most of recorded time. Its striking appearance makes it noticeable to anyone who bothers to look up. This is because, with the exception of the Moon, Venus is brighter than any other object in the night sky.

Since Venus is closer to the Sun than Earth is, the planet's orbit never takes it far from the Sun (when viewed from our planet). As a result, Venus can only be seen in the early morning sky, rising in the east just before sunrise, or in the early evening sky, sinking toward the western horizon just after sunset.

What is the Evening Star?
Venus

The planet Venus is so bright, it has not only been mistaken for a "Morning Star," it has also been mistaken for an "Evening Star"—when its orbit takes it to the other side of the Sun and into Earth's evening skies. While above the western horizon, Venus will dominate the evening sky as a brilliant pinpoint of light.

How do they name stars?
The brightest stars were all named long ago. The fainter stars are known by various designations that usually correspond to what portion of the sky the star lies in. For any new discoveries, the International Astronomical Union is responsible for assigning names.

There are many organizations that are willing to sell you a star, galaxy, or other astronomical object that you can then "name." While these companies are perfectly legal, the truth is that there is only one scientifically recognized authority that names astronomical objects—the International Astronomical Union (IAU)—and they do not sell star names.[3]

Composed of professional astronomers from around the world, the IAU has established specific guidelines when it comes to the naming of different types of celestial objects. However, most of these guidelines refer to solar system objects (such as hills on Mars, or a newly discovered asteroid). Since stars are constant and have been around since the dawn of time, traditional names for the brighter stars were already in place long before the IAU was established in 1919.[4] In addition, various stellar catalogs and a coordinate system had already been established to help identify the stars. So, even before the IAU, many of the stars in the night sky were identified several different ways. For example, the bright star Rigel in the constellation of Orion is also known as Algebar, Elgebar, Beta Orionis, SAO 131907, GSC 5331:1752, HIP 24436, PPM 187839, HD 34085, and RA 051432, Dec –008120. To avoid confusion, most astronomers today—with the endorsement of the IAU—simply refer to stars by their catalog numbers or astronomical coordinates.

Companies that want to add your name to the confusion by selling you a star are not recognized by the IAU.[5] Buying one of these stars is similar to buying real estate on another planet. No one really has the authority to make such a sale. If you have been thinking about buying a star for a friend or family member, why not take her to a planetarium or a public observatory instead, where she can learn about the real thing?

Do stars really fall?
No.

A so-called falling star has nothing to do with the stars you see at night. The real nighttime stars are all huge balls of hydrogen and helium gas that are light-years away from us. A falling star is only a fragment of rock that runs into Earth's atmosphere. As noted earlier, that tiny piece of rock is called a meteor.

When a dust particle or a rock hits Earth's atmosphere, friction between the two causes the rock to burn up. The hot rock leaves a bright streak across the sky. Most of the time, these small rocks and dust particles burn up completely. Other times, some of the rock survives to land on Earth. When it hits the ground, the rock becomes known as a meteorite.

Why do stars come out only at night?
Because light from the Sun drowns them out during the day.

Our planet is surrounded by stars all the time, even in the brightest part of the day. However, because the closest star, the Sun, is so much brighter than others in our skies, its brilliant light drowns out the much fainter stars during daylight hours. It is only when our side of Earth turns away from the Sun that we can see the other stars.

Like just about everything, there is an exception to this rule. During a solar eclipse, when the Moon completely covers the disk of the Sun, the sky gets dark enough for us to see a few bright stars during the day. They are visible only for a few minutes. The moment a sliver of the Sun's disk peaks around the Moon, the sky brightens and the other stars disappear.

How many stars can you see with your eyes alone?
About five thousand stars.

There are literally billions of stars in our Milky Way galaxy. But even on the darkest, clearest night, the human eye can only pick out about five thousand individual stars—and that's only if you stay up all night *and* could see the stars that the Sun blocked during the day![6] Human eyes are not sensitive enough to see any of the other billions of stars. They are too far away and too faint to see unless you use a telescope.

Do people in different parts of the world see the same stars at the same time?
No but yes.

Because Earth is shaped like a rotating ball, half of the planet is experiencing daylight while the other half is experiencing night. As a result, there is no way that everyone around the world can see the same stars at the exact same time. However, within a twenty-four-hour period, people living at the same latitudes around the globe will see the same stars at the same local time. In other words, people living in the northern United States and Canada will see the same stars at 10:00 PM local time that people in Europe will see at 10:00 PM their time. People living in Australia will see the same stars at 10:00 PM as people living in southern Africa and central South America at 10:00 PM their time. However, people in the United States will not see the same stars at 10:00 PM as people in Australia.

What a person sees in the night sky depends on the latitude at which he lives. People who live along the equator will see all of the same stars, the stars of both the Northern and the Southern Hemispheres. People living at the North Pole would see stars only of the Northern Hemisphere, while those at the South Pole would see only stars of the Southern Hemisphere. People living in the middle latitudes of the Northern Hemisphere would see all the stars of the Northern Hemisphere and a few from the Southern and vice versa.

Do stars come in different colors?
Yes.

Even though most of the stars in the night sky appear as white pinpoints of light, this impression has more to do with the poor night vision of humans rather than the stars themselves. In actuality, stars come in many different colors, including red, yellow, blue, and white. The reason we see mostly white stars is because the sensors in our eyes responsible for detecting color do not work well in the dark. However, if you look carefully at the night sky, you will be able to detect slight hints of color from some of the brighter stars. For example, Betelguese, in Orion, will have a reddish twinkle to it. (This star is a red supergiant star.) Rigel, also in Orion, will have a slight blue cast to it. (Rigel is a blue giant star.) To get a much better sense of the colors of stars requires a long-exposure photograph. Far more sensitive than the human eye, a photograph reveals the true colorful nature of the stars. (Look carefully at the astronomical images in this book's color insert for examples.)

A star's color is dependent upon how hot the star is burning. Red stars, which are the "coolest" type of stars, have a surface temperature of 7,000°F (4,000°C). Yellow stars like our Sun have a surface temperature of approximately 10,000°F (6,000°C), while blue stars are the hottest with surface temperatures reaching as high as 63,000°F (35,000°C).[7] So much for our Earthly perceptions of hot and cool colors!

What is a constellation?
A collection of stars that we have formed into an imaginary pattern,
like Leo, the Lion, or Orion, the Hunter.

A long time ago, people played connect-the-dots with the stars and envisioned patterns in the sky. These ancient patterns were passed down through the generations and have become the constellations we know today.

Most people are familiar with the Big Dipper, a group of stars that is part of the constellation Ursa Major (the Big Bear).

The seven stars of the Big Dipper are connected only by our imagination. These stars actually vary greatly in their brightness and distances from Earth. From Earth, they appear to form a dipper. From a different position—say, from a planet orbiting another star—the seven stars would form a different shape. We would be seeing them from a completely different perspective.

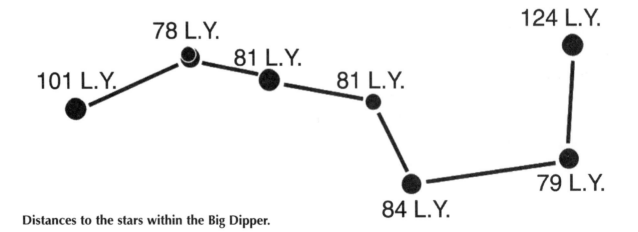

Distances to the stars within the Big Dipper.

How many constellations are there?
Eighty-eight.

There are a total of eighty-eight constellations in the sky. This includes all of the constellations in the Northern and the Southern Hemispheres. (A list of all eighty-eight constellations can be found in the Quick Facts section on page 216.)

People living in the Northern Hemisphere will not be able to see all eighty-eight constellations. Most of the southern constellations will be below the horizon, hidden from view. The reverse goes for people living in the Southern Hemisphere. They won't be able to see many of the constellations to the north.

Do astronomers study constellations?
No.

Constellations are great for finding your way around the sky and for telling stories around a campfire. In the old days, sailors like Christopher Columbus and Ferdinand Magellan used constellations to find their way on the high seas. Today, however, astronomers tend to use computer programs and coordinate systems to find stars of interest.

Constellations are huge imaginary figures containing many of the bright stars in the sky. Astronomers study individual stars and physical groups of stars such as star clusters and galaxies. They study objects in the sky, not imagined patterns of stars on the sky. Also, astronomers have divided the sky into a coordinate system, so they no longer need constellations to help them find the fainter stars.

Can you see a constellation through a telescope?
No.

When you look through a telescope, you see only a few stars in a small area of the sky. Stars in a constellation are stretched across a large portion of the sky. Even the smallest constellations are far too large to view in a telescope.

How can I find a constellation?
Start with a few bright stars or a well-known constellation.
Then use those to help you find others.
The Big Dipper in Ursa Major and Orion, the Hunter, are good places to start.

The following two star tours will demonstrate how you can find several constellations just by learning one or two.

Quick Spring/Summer Sky Tour

In the summer months, the Big Dipper is high overhead in the Northern Hemisphere and easily recognizable, so we will start there. The Big Dipper is just a nickname for the seven brightest stars of the constellation Ursa Major.

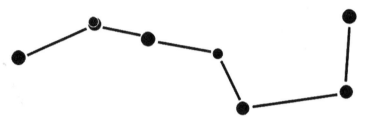

Four stars make up the scoop of the dipper, while three stars make up its handle. Two stars in the scoop are called the pointer stars because they point to a star called Polaris. Polaris is also known as the North Star. You will always find Polaris in the north.

The North Star is the last star in the handle of the Little Dipper, or Ursa Minor (the Little Bear). The Little Dipper is much smaller and much fainter than the Big Dipper.

Now, go back to the Big Dipper and visually connect the stars in the handle.

Do the stars in the handle form a line? No. Instead, they form a curve or an arc. If you follow that curve away from the Big Dipper, the arc it forms leads to a star called Arcturus. So you follow the arc of the handle to Arcturus, "arc to Arcturus."

Arcturus is the brightest star in the constellation of Boötes (pronounced boo-a-tees). Boötes is supposed to be a shepherd, but it looks more like a kite or an ice cream cone.

Now, let's go back to the Big Dipper's handle. If you continue to follow the arc formed by the

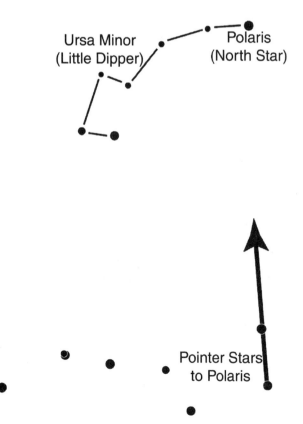

Ursa Minor (Little Dipper)

Polaris (North Star)

Pointer Stars to Polaris

handle of the dipper, you will not only find the "arc to Arcturus," you will also "speed on to Spica."

Spica is the brightest star in the constellation of Virgo. In mythology, Virgo was the Goddess of Earth, protector of all the creatures on its surface. In the sky, Virgo is made up of a group of faint stars, with the exception of bright Spica.

You have now learned how to find four constellations—Ursa Major, Ursa Minor, Boötes, and Virgo—just by starting with the seven stars of the Big Dipper.

Quick Winter Sky Tour

For a tour of the winter sky, start with Orion, the Hunter. Orion is visible from both the Northern and the Southern Hemispheres and has a distinct pattern of three stars that form what is called Orion's belt.

You can use the belt stars to point to a couple of different constellations. If you follow the belt stars to the west, you will find a small group of stars that look like a tiny dipper. These are the Seven Sisters or the Pleiades in the constellation of Taurus, the Bull. One story has Orion chasing the Seven Sisters around the sky.

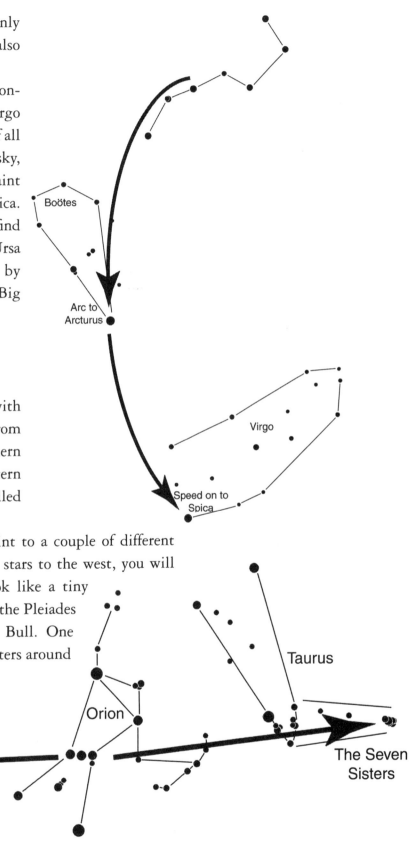

If you follow Orion's belt stars to the east, you will find Sirius, the Dog Star, in the constellation Canis Major, the Big Dog. Many consider Canis Major to be Orion's hunting dog. Sirius is the brightest star in the night sky. So by learning one constellation—Orion—you can find two more, Canis Major and Taurus.

There are other ways to group stars together so you can find them. Start slow, with a couple of constellations at a time, and gradually you will become familiar with the night sky.

What are the constellations of the zodiac?
Twelve constellations that lie along the ecliptic—
the path the Sun appears to follow across the sky.

Even though it is our planet that is moving around the Sun, from our vantage point on Earth it looks as if the Sun slowly moves across the sky with respect to the background stars. Since Earth always follows the same path around the Sun every year, the Sun seems to travel through the same constellations at the same time each year. The constellations that the Sun appears to travel through are known as the constellations of the zodiac and include Aries, Taurus, Gemini, Cancer, Leo, Virgo, Libra, Scorpius, Sagittarius, Capricorn, Aquarius, and Pisces.

To help explain this movement, look around the room and find a light source that you can use to represent the Sun, preferably a lamp at about eye level. Now, imagine that you are Earth. Look at the "Sun" and then notice what objects lie to its right and left. These objects would represent constellations of the zodiac. Now, move a little with respect to the light. Even though you are the one moving, it will appear that the "Sun" has changed position with respect to these constellations.

To notice this effect in the real world, find a spot to watch the Sun go down. (Be careful not to look directly at the Sun since it is dangerous to your eyes.) Notice its exact position along the horizon when it disappears. Then, as the sky darkens, you will be able to make out one of the constellations of the zodiac just above this point. Repeat this observation in a month. You will see a different constellation of the zodiac. There will be a new constellation visible each month for a year, then the pattern repeats itself.

How do astronomers measure the distances to stars?
They use parallax, or a combination of a star's apparent brightness
and its luminosity, to determine distance.
They also use variable stars, supernovae, and the redshift of distant galaxies.

(Prepare yourself. This simple question has a long, rather complicated answer.)

The most reliable method for measuring distances to nearby stars is parallax. When using this method, a star's location is carefully noted with respect to the surrounding background stars. Then, six months later, when Earth is on the opposite side of the Sun, the same star is observed again. If the star's position has moved however slightly with respect to the background stars, this means that it is relatively close by.

Astronomers then measure the angular distance between the star's two observed positions. With this angle and the distance between the two observing locations (which is equal to about 2 AUs because Earth was on opposite sides of the Sun when the two readings were taken), astronomers can form a triangle and use a simple trigonometric equation to determine the distance to the star. (For those mathematically inclined individuals, the equation is $d = 1\ \text{AU}/\tan(\theta/2)$ where d = distance to star and θ = angular distance between star locations.)

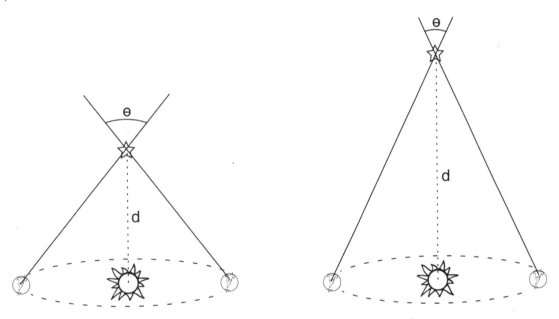

Using parallax to determine distance to nearby stars.

The European Space Agency's *Hipparcos* satellite measured the stellar parallax of almost 120,000 stars at distances as far out as several hundred light-years.[8] Beyond this distance, the apparent movement of stars is too slight to be detected and astronomers must look to a star's spectrum for hints of its distance—a method known as spectroscopic parallax. Though this method involves a spectroscope, it has nothing to do with parallax. (A spectroscope is an instrument that takes light from a star and breaks it up into its component elements.)

When light from a star passes through a spectroscope, it travels through a prism or diffraction grating and is spread out into its individual colors, forming a spectrum. The temperature and luminosity (total energy output) of a star can be determined by the appearance and intensity of certain elements within its spectrum.[9] This information, along with the star's apparent brightness (easily measured with a photometer), is used to determine the star's distance. (Again, for the mathematically inclined, the equation is: $L = b4\pi d^2$, where L = luminosity, b = apparent brightness, and d = distance.)[10]

Unfortunately, there are some inherent uncertainties in calculating the star's luminosity from its spectrum. As a result, the distance calculations made using the spectroscopic parallax method are accurate only to within about 10 percent.[11] Even with this rather large uncertainty, the measurement gives astronomers a general idea of the star's distance, which is more than they would have otherwise.

Another method used to determine stellar distances involves specific types of variable stars. A variable star is one whose luminosity varies over time. For certain categories of variable stars, this change occurs in a repeatable pattern. In the case of both Cepheid variables and RR Lyrae variable stars, their average luminosities appear to be tied directly to their regular periods (the time it takes one of these stars to cycle from faint to bright and back to faint again). The shorter the period, the lower the star's average luminosity. The longer the period, the greater its luminosity.

Astronomers have created a formula to explain this relationship between period and luminosity.[12] Now, when they observe the period of a Cepheid or RR Lyrae variable star, all they have to do is plug the length of the period into their formula and out comes the star's average luminosity. Knowing a star's average luminosity and apparent brightness (which is measured separately), astronomers can then calculate its distance using the equation from above ($L = b4\pi d^2$). Because Cepheid variables are so bright, astronomers have been able to observe these stars in galaxies millions of light-years away.

While Cepheid variable stars are great for determining distances, there are many regions of space that don't include these wonderfully predictable stars. In addition, the distances in space are so vast that even Cepheid variables become too faint to be seen after a while. So astronomers must come up with even more methods for measuring distances, including observations of Type Ia supernovae. (The causes and types of supernovae are discussed in detail beginning on page 150.)

Astronomers believe that a Type Ia supernova has a definite limit to its maximum luminosity. If this is indeed the case, each Type Ia supernova at its peak would have the same total luminosity.[13] So when one of these supernovae is observed, astronomers simply measure its apparent brightness, then use the established maximum luminosity value and the formula from above ($L = b4\pi d^2$) to determine the distance to the exploding star. Supernovae are much brighter than even the brightest Cepheid variable stars, allowing astronomers to see even farther into space. However, supernovae are rare events. Since astronomers never know where or when one is going to take place, they have to take advantage of one as it happens.

Despite the brightness of these supernovae, there are galaxies in our universe that are so far away that astronomers are unable to see even the brightest supernova explosions within their boundaries. At this point it becomes necessary to invent yet another means of measuring distances. In 1929 an American astronomer by the name of Edwin Hubble (after whom the Hubble Space Telescope is named) created an equation—now known as Hubble's law—which allows astronomers to determine the distances to extremely distant galaxies simply by measuring the amount of redshift in their spectra.[14]

When an object is moving away from us at tremendous speeds, everything within its spectrum is shifted toward its red end by the Doppler effect. (Everything maintains its proportional spacing, it all just gets shifted toward the red side.) Hubble observed that the more distant a galaxy, the faster it was traveling away from us and the greater its redshift. It was as if the universe was expanding, getting faster as it moves farther out. He calculated a rate at which these galaxies were expanding, a rate that became known as Hubble's constant. With this constant and the difference in wavelengths caused by the redshift, any astronomer could determine the distance to these far-off galaxies.

Hubble's equation is: $v = H_0 d$ where v = the receding velocity of the galaxy, H_0 = Hubble's constant, and d = distance.[15]

Since it seems that these extremely distant galaxies lie at the very fringes of our universe, it is doubtful there is much more beyond them that we will have to measure. So, we will cease our measuring discussion for now, thus bringing to a close the rather lengthy answer to the simple question "How do astronomers measure the distances to stars?"

What is a light-year?
A unit of measure equal to 5.9 trillion miles (9.5 trillion kilometers), the distance that light travels in one year.

In astronomy, distances are so great that astronomers had to invent a new unit to describe a distance of trillions of miles. If they didn't, they would have to spend most of their time writing zeroes. The unit they created was the light-year. One light-year is equal to the distance that light travels in a year. You may not realize that light takes time to travel anywhere. In fact, scientists believe that nothing can travel faster than the speed of light. Light travels 186,000 miles (300,000 kilometers) every second. If you lived 186,000 miles from the nearest power station, it would take one second for the light to reach your house after you hit the switch.

To determine how far light travels in a year, it is necessary to multiply 186,000 miles per second by 60 seconds in a minute, 60 minutes in an hour, 24 hours in a day, and 365 days in a year. After multiplying all of these together, you would get:

$$186{,}000 \times 60 \times 60 \times 24 \times 365 = 5{,}900{,}000{,}000{,}000 \text{ miles/year}$$

or

$$300{,}000 \times 60 \times 60 \times 24 \times 365 = 9{,}500{,}000{,}000{,}000 \text{ kilometers/year}$$

Even traveling that fast, it still takes time for light to travel through the huge distances in space. Within our solar system, light takes 8.5 minutes to travel from the Sun to Earth. To get to Pluto, it takes light 5.5 hours.

Light traveling from our Sun to the nearest star system, Alpha Centauri, takes 4.2 years to get there. As a result, astronomers record the distance to Alpha Centauri as 4.2 light-years. As you can see, this is much easier than writing Alpha Centauri's distance as 24,600,000,000,000 miles (39,700,000,000,000 kilometers). Keep in mind that this is only the closest star. All of the other stars are much farther away. (For a listing of the twenty-five closest stars to us with their distances listed in light-years, please see page 214 of the Quick Facts section.)

What is a parsec?

A unit of measure equal to 3.26 light-years, the distance at which two objects that are one AU apart are separated by one arc second.

When astronomers were first using the parallax method to determine the distances to nearby stars, they wanted to create another unit of measure beyond the light-year, something based on the parallax techniques they were using in their calculations. What they came up with was the parsec, short for *par*allax of one arc *sec*ond. How they ended up with a value of 3.26 light-years will take a little explaining.

First, remember that the distance between Earth and the Sun is equal to one AU. Now, imagine traveling away from Earth and looking back at the Earth/Sun pair that is still separated by one AU. The distance between the two hasn't physically changed, but the pair's angle of separation has. It's as if the two are moving closer together as you move farther away from them. If you keep moving away, the pair's angle of separation will continue to decrease. A parsec is the distance you need to travel for the angle of separation between Earth and the Sun to equal one arc second.

One arc second is a very small angle, equal to 1/3600 of a degree. To help picture just how small that is, think of the last Full Moon you saw. A Full Moon covers 30 arc minutes in our sky. So, divide the Full Moon into 30 arc minutes. Then, divide each one of those arc minutes into 60 arc seconds. The angle you are looking for is one of those arc seconds. Astronomers calculate that you would have to travel 3.26 light-years from our solar system before the angle of separation between Earth and the Sun is small enough to equal one arc second.

Astronomers didn't stop with just the parsec. They also use kiloparsec (1,000 parsecs) and megaparsec (1,000,000 parsecs) to describe the vast distances between galaxies. However, there really isn't that much difference between a light-year and a parsec. Is one better than the other? No. It's just that some astronomers use parsecs and others use light-years.

What is the closest star to Earth?
The Sun, but most people forget that the Sun is a star.

The next closest star to Earth is called Alpha Centauri, the brightest star in the southern con-stellation Centaurus. As noted earlier, this star system is 4.2 light-years, or 24.6 trillion miles (39.7 trillion kilometers), away from us. The star lies too far to the south to be seen from most of the Northern Hemisphere. While Alpha Centauri is the closest star in the night sky, it is not the brightest. Two other stars, Sirius and Canopus, are farther away but brighter than Alpha Centauri.

A telescope reveals that Alpha Centauri is actually a triple star system.[16] The two larger stars of the system orbit in a tight circle around each other and have been given the names Rigel Kent A and Rigel Kent B. A third, fainter star, called Proxima Centauri, is actually somewhat closer to us and orbits around the brighter pair. (For a list of the twenty-five closest stars, see page 214 of the Quick Facts section.)

How does a star form?
Within a giant nebula of gas and dust, gravity will slowly attract
more and more gas and dust together until it has formed a giant sphere.
As gravity continues to attract mass to the sphere, pressures within its core increase.
The temperature increases as well. The pressure and temperature will
continue to rise until they are high enough for nuclear fusion to begin.
At that point, a star is formed.

Stars begin their lives in giant clouds of gas and dust known as diffuse nebulae. These neb-ulae are made up mostly of hydrogen gas with a scattering of dust particles throughout. The gas and dust are not evenly distributed within the cloud. Instead, they clump together in places. (If you were wondering, "clump" can be considered a technical term!)

Two views of the beautiful diffuse nebula known as the Eagle Nebula
can be seen in the color insert section of this book.

These clumps have more mass than the individual atoms of gas and specks of dust in the surrounding area. Since more mass means more gravity, the smaller particles are gravitation-ally attracted to the larger clump. As more matter is drawn in, these clumps become more massive and in turn attract still more matter to them.

While a clump grows in size, the gas in its center is being squeezed by the strong gravity. As the gravity squeezes the hydrogen atoms together, the temperature and pressure in the core of the clump begin to rise. It's like being in a small room. The more people that crowd into the room, the hotter the room gets.

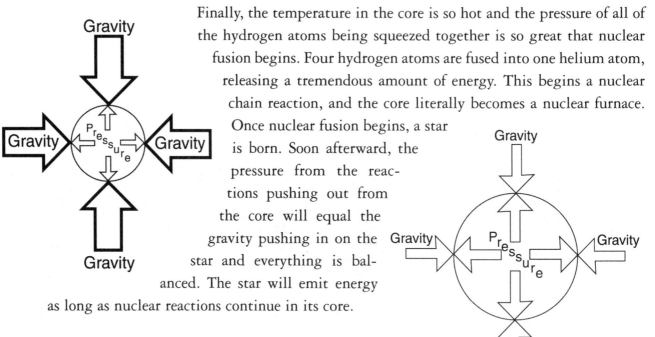

Finally, the temperature in the core is so hot and the pressure of all of the hydrogen atoms being squeezed together is so great that nuclear fusion begins. Four hydrogen atoms are fused into one helium atom, releasing a tremendous amount of energy. This begins a nuclear chain reaction, and the core literally becomes a nuclear furnace.

Once nuclear fusion begins, a star is born. Soon afterward, the pressure from the reactions pushing out from the core will equal the gravity pushing in on the star and everything is balanced. The star will emit energy as long as nuclear reactions continue in its core.

Will a star live forever?
No.

In some ways, stars are just like humans. They are born, live out their lives, and then die. But instead of living for seventy, eighty, even ninety years, like some humans, they live for millions or billions of years.

After living an extremely long life (by human standards), a star will run out of fuel and die. If the star is small, it uses up its fuel gradually and slowly fades away. A star like our Sun will go through a couple of expansions and contractions before it exhausts all of its fuel. (For more details, see "Will the Sun just burn out and go black?" on page 36.) If the star is many times larger than our Sun, it will use all of its fuel quickly and explode in a supernova. Nothing lives forever.

How long do stars live?
Millions, billions, or trillions of years, depending on their size.

When a star is born, its size determines how long it will live. The smaller the star, the longer it will live. The larger the star, the quicker it will die.

The small, low-mass stars fuse hydrogen into helium very slowly. These cool, red stars (known as red dwarfs) will burn for trillions of years before they use all the hydrogen in their cores. When all of their fuel is exhausted, these stars will simply burn out with little fanfare.

Medium-size stars (like our Sun) burn faster. Because they are more massive than the red stars, the pressures in their cores are greater, causing nuclear reactions to occur more quickly. Medium-size stars live for a few billion years. In their last stages of life, these stars become red giants, form planetary nebulae, and wind up as white dwarfs before fading to black.

The larger, more massive stars have the shortest lives. These stars (known as blue giants) are the hottest and brightest in our skies and will use up all of their fuel in only a few million

The Helix Nebula (also known as NGC 7293) is a planetary nebula, a cloud of gas and dust surrounding a dying star. Image credit: Jack Newton, Arizona Sky Village

years. As their fuel supply runs short, these stars will swell to become red supergiant stars. Then, their final collapse results in dramatic supernova explosions. These stars leave behind exotic entities like black holes, neutron stars, and pulsars.

So when it comes to stars, the bigger you are, the shorter and more dramatic your life.

What is a binary star?
Two stars in orbit around each other.

A binary star system, or double star, contains two stars that orbit around a common point. Our Sun is not part of a binary star system, which places it in the minority. Over two-thirds of all the stars in the sky have at least one companion star. Some star systems contain three or four stars. The light from Castor, a bright "star" in the constellation of Gemini, is actually made up of the light from six individual stars.[17]

Binary star systems include stars that are bound together by gravity. Visual binaries are two stars that only look as if they are connected. Mizar, the middle star in the Big Dipper's handle, is part of a visual binary star system. Slightly above and off to one side of Mizar lies a fainter star called Alcor. The two stars look as if they are a pair, but they have no gravitational ties to each other. The bizarre truth of the matter is that Mizar is a binary star, and so is Alcor.[18] They each have a companion star that is too faint to be seen by the human eye. So in this case, we can see a visual binary system, but can't see the real binary pairs.

What is an open cluster?
A young group of stars recently formed from the same diffuse nebula.

Stars are formed from within a diffuse nebula, an enormous cloud of gas and dust. No matter how large the nebula is, the star-forming process will eventually use up all of its material. When there isn't enough gas and dust left to form another star, all that will remain of the original gas cloud is a loose grouping of young stars. These stars will be of different sizes and luminosities, but all will be approximately the same age—having just formed from the same nebula. This group of young stars is known as an open cluster. One of the most famous open clusters is called the Pleiades, or Seven Sisters, located in the constellation of Taurus.

In the life cycle of a star, a diffuse nebula could be considered a stellar nursery and an open cluster could be considered a stellar kindergarten. While several of the stars will group together to form multiple star systems, the group as a whole will gradually disperse. After a few million years, there will be little evidence of a nebula or an open cluster in this region of space.

The Pleiades (or Seven Sisters) located in the constellation of Taurus, is an example of an open (or galactic) star cluster. Image credit: NASA, ESA, and AURA/Caltech

What is a galactic cluster?
Another name for an open cluster.

Astronomers sometimes use the term "galactic cluster" to describe the loose grouping of young stars explained in the previous answer. "Galactic" refers to the fact that all open clusters are found within a galaxy. This helps to distinguish these clusters from another type of star cluster, a globular cluster, whose members are found only on the outskirts of a galaxy.

What is a globular cluster?
A large, tightly knit sphere of old stars that lies on the outskirts of a galaxy.

When viewed through a telescope, a globular cluster looks like a delicate pile of sugar crystals. Each of these clusters is made up of thousands of stars that are bound together by gravity to form a giant glowing sphere several light-years across. The stars that make up the globular clusters have been around for a while, probably since around the formation of the galaxy itself. Astronomers aren't exactly sure how these clusters formed, but whatever forces are involved must be common throughout the universe since globular clusters have been found surrounding other galaxies like our own.

M80 (NGC 6093) is a beautiful example of a globular cluster. Image credit: Hubble Heritage Team (NASA/AURA/STScI)

What is a brown dwarf?
An object that is too small to be a star, but too large to be a planet.

Sometimes found in orbit around other stars, sometimes found on their own, these large, dark balls of gas are at least thirteen to eighty times more massive than Jupiter—too massive to be considered a planet.[19] However, brown dwarfs are not massive enough for nuclear reactions to take place in their cores, so they are not stars. They fall somewhere in between.

What is a red giant?
A big, old, red star.

Stars are not born as red giants. Instead, a star like our Sun becomes a red giant when it has used up most of its fuel and begins to grow in size. As a star begins to swell in size, its surface cools off and turns red in color. Hence, the "red" part of the term. (When it comes to star colors and temperatures, cool = red, warm = yellow, and hot = blue/white.)

When our Sun becomes a red giant, a few billion years from now, it will swell in size until it swallows Mercury and Venus and extends almost all the way out to Earth. Hence, the "giant" part of the term.

What is a red supergiant?
A really huge, old, red star.

These incredibly huge stars begin their lives as blue giants, the largest type of star formed in a diffuse nebula. Because it is so large, the pressure and temperature in the core of a blue giant are tremendous. Under such conditions, this star uses up its fuel quickly. When its fuel supply is almost exhausted, a blue giant star begins to swell in size in an attempt to maintain the equilibrium between its pressure and gravity. As it swells, its surface cools and becomes red.

Betelgeuse, a star in the constellation of Orion, is an example of a red supergiant star. If Betelgeuse were to replace our Sun, this star would completely swallow Mercury, Venus, Earth, and Mars, and its surface would reach almost out to Jupiter. In other words, Betelgeuse would completely fill our entire inner solar system.

A star will not stay at the red supergiant stage for very long. After a few thousand years or so, it will use up the rest of its fuel and explode in a violent supernova. The remnants will form either a pulsar, a neutron star, or a black hole.

A stunning image of red supergiant V838 Monocerotis can be found in the color insert section of this book.

What is a white dwarf?
A "dead" star about the size of Earth.

After a star like our Sun uses up all of its fuel, it will collapse into a small ball about the size of Earth. The resulting small, dense sphere is called a white dwarf. A white dwarf remains very hot and bright for a long time as it cools off, but no new nuclear reactions occur in its core. Eventually it will cool completely and fade to black.

Within a white dwarf, matter that was once spread out inside an entire star is now compressed into a sphere the size of a small planet. This makes the star so dense that a teaspoon of material from a white dwarf would weigh a couple of tons!

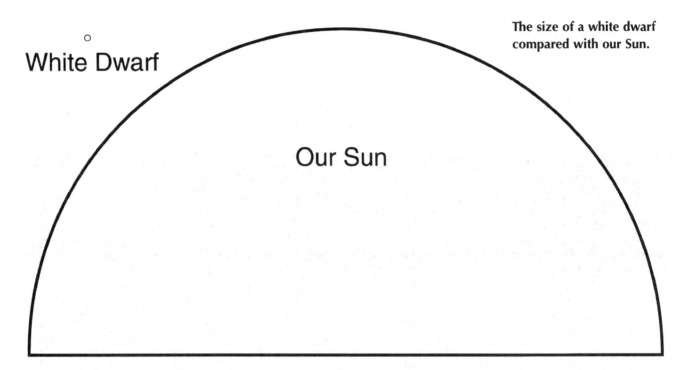

White Dwarf

Our Sun

The size of a white dwarf compared with our Sun.

What is a nova?
A brief, powerful explosion on the surface of a white dwarf star.

As discussed above, a white dwarf is a dead star. So how can a dead star explode? With help. If a white dwarf star is located in a binary star system where the orbit of the two stars brings them fairly close together, sometimes the strong gravity of the white dwarf will attract hydrogen from the other star. This stolen hydrogen will slowly accumulate on the surface of the white dwarf, where it will be compressed by the strong gravity. As more hydrogen is collected and compressed, the temperature on the surface of the white dwarf begins to rise. Eventually, the temperature and pressure are great enough for nuclear fusion to begin. In a brief burst, all the "stolen" hydrogen is used up, and the reactions cease. This brief bout of nuclear reactions produces a bright flash of light known as a nova.

What is a supernova?
A huge explosion that rips apart a dead or dying star.

Astronomers believe there are two different types of supernova explosions.

A Type Ia supernova occurs when a white dwarf is trapped in a close orbit around a red giant star. The gravity from the white dwarf will attract hydrogen from the other star's atmosphere. As the red giant swells in size, its surface gets closer and closer to the white dwarf, allowing large quantities of hydrogen to pass quickly to the small, dense star. This rapid increase in mass is what proves explosive for the white dwarf. The rapidly increasing mass causes a dramatic increase in the pressure and temperature at the star's core. Very quickly, conditions in the core reach a point where the solid carbon core can ignite and nuclear fusion begins.

Because matter in a white dwarf is packed in so tightly, it is unable to expand once this new round of nuclear reactions begins. Tightly confined, the reactions cause the temperature

The Crab Nebula was formed when a supermassive star went supernova in the year 1054.
Image credit: Jack Newton, Arizona Sky Village

and pressure to increase even more rapidly until the star literally blows itself apart. Because there are physical limits to the size of a white dwarf, astronomers know how much additional matter would be required to ignite the nuclear reactions in the core. Therefore, they can determine the forces of the supernova explosion and predict how bright it will be. Knowing the total luminosity of the explosion and its apparent brightness allows them to measure the distance to the exploding star.

A Type II supernova occurs with the death of a very large star. When a supermassive star—one that is over three times more massive than the Sun—uses up its fuel, the forces from the nuclear reactions that have been pushing outward from the core suddenly disappear when the reactions stop. The gravity pushing inward immediately takes over and causes the star to collapse in on itself. This collapse happens so fast that the star's material collides with itself in the center. A tremendous rebound occurs and the star is torn apart in a resulting supernova.

A supernova releases so much energy that the star's brightness can increase one hundred million–fold. A star ordinarily not visible from Earth can suddenly become the brightest star in our sky and can even be seen during the day. Chinese astronomers recorded just such an event in the year 1054.[20] The resulting debris from this stellar explosion can still be seen today with the aid of a telescope. Located in the constellation Taurus near one tip of the Bull's horns, this expanding supernova remnant (the stellar remains) is known as the Crab Nebula.

Just how bright a Type II supernova will become is extremely hard to predict. Before a star can go supernova, it must have a starting mass of at least four times that of our Sun.[21] This is just a lower limit. There are stars that are almost thirty-five times more massive than our Sun. With such a wide range of masses, it is impossible for astronomers to predict the total luminosity of a Type II supernova. It just a safe bet to say that they will get really bright.

What is a neutron star?
A very small, dense remnant of a dead star.

A neutron star is considered "dead" because no nuclear reactions are occurring in its core. This type of star forms after a supernova explosion. You might ask: if a supernova occurs when a huge star rips itself up, how can anything form after such a tremendous explosion? Sometimes when a supernova explosion takes place, the core of the star survives. With the rest of the star destroyed, the core begins to collapse in on itself. The leftover material collapses into a sphere about ten miles across. This is a neutron star.

A neutron star is even denser than a white dwarf. A teaspoon of material from a neutron star would weigh several million tons. Think about this. Much of the material that was in a star many times larger than our Sun is now crammed together inside a ball smaller than the state of Rhode Island. That's dense!

What is a pulsar?

A rotating neutron star. As it spins, a pulsar seems to be sending a series of radio pulses to Earth. The word "pulsar" is short for pulsating star.

When astronomers discovered the first pulsar in 1967, they had no idea what it was. They found that if they aimed a radio telescope at a specific spot in the sky, they received a series of "blips." These radio blips were so regular that the astronomers didn't believe they could be produced by anything natural. They called the unknown object "LGM," short for Little Green Men.[22]

As you may have guessed, it turns out that pulsars don't have anything to do with little green men. Instead, a pulsar is a rapidly spinning neutron star located within the debris left over from a supernova—an expanding cloud sometimes referred to as a supernova remnant. The pulses are caused by its fast rotation, the neutron star's spin axis not pointing directly at Earth, and its magnetic pole not lined up precisely with its spin axis.

Before we tackle the problem of the pulsar, let's examine something more familiar. Imagine a friend has a flashlight pointed toward you. Now your friend begins to spin around. You see a flash of light every time your friend turns the flashlight in your direction. When your friend is facing away from you, you don't see anything.

Now imagine your friend is a pulsar and the flashlight he's holding is the pulsar's magnetic north pole. Pulsars have extremely strong magnetic fields, much stronger than Earth's. Charged particles trapped in this field release radiation that escapes the pulsar at its magnetic north and magnetic south poles. We detect this radiation as a blip every time the north or south magnetic pole points in our direc-

tion, just as you detect the light every time your friend turns the flashlight toward you.

There is something else you can learn about pulsars. Let's go back to your friend with the flashlight. You can tell how fast your friend is spinning by counting how long it is between flashes of light. The faster he spins, the faster you will see the blips of light. Your friend may be able to spin around once every one or two sec-

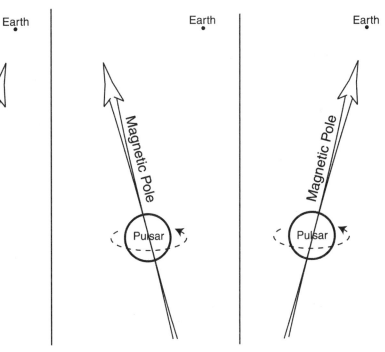

onds, but he will probably get dizzy and have to stop. Pulsars have been found to rotate on their axes anywhere from once every second to thirty times a second.

What is a quasar?
A quasi-stellar radio source that is moving away from us at tremendous speeds.

The term quasar is short for quasi-stellar radio source. Although starlike in appearance, quasars have nothing in common with a typical star. Visually, quasars are fairly faint, yet they produce some of the strongest radio sources in the universe. In addition, quasars appear to be fairly small, especially when compared to a common galaxy. The constant stream of intense radio emissions coming from such a small source puzzles astronomers.

Another mystery surrounding quasars is their speed. For some reason, quasars are the fastest moving objects in the universe. A typical quasar spectrum shows a large amount of redshift (where all of the typical element lines are shifted toward the red end of the spectrum). This large redshift indicates that these objects are moving away from us at extremely high velocities. As a result, quasars are some of the most distant objects known.

It has been suggested that quasars are extremely young galaxies that are dominated by supermassive black holes.[23] However, because of their great distances and tremendous speeds, it may be impossible to ever know for sure.

ABOUT BLACK HOLES

What is a black hole?

What is a singularity?

What is an event horizon?

How do black holes form?

How black is a black hole?

Since black holes are "black," how do you find them?

What would happen if our Sun became a black hole?

Could a human travel through a black hole?

Can we send a spaceship through a black hole?

Where does everything go after it enters a black hole?

Will black holes devour the universe?

What is a black hole?
A place where gravity is so strong that nothing—
not even light—can escape its clutches.

For us earthlings, gravity is a good thing. It is what keeps us and everything else attached to our planet. For example, if you throw a ball up in the air, gravity will pull the ball down until it hits the ground. The faster you throw the ball, the farther it will travel before gravity pulls it back to Earth. If, however, you could throw the ball really fast, faster than 25,039 miles per hour (40,320 kilometers per hour)—Earth's escape velocity—it would be able to escape Earth's gravitational pull and leave our planet completely. (Keep in mind that the best pitchers in baseball can only throw a ball about 100 miles per hour, or 160 kilometers per hour.) In other words, leaving our planet is no easy feat, but we can do it with the help of some incredibly powerful rockets.

Leaving a planet or other body that is more massive than Earth is even more difficult. Because gravity is directly related to mass, the more massive the object, the stronger its gravity. The planet Jupiter is almost 319 times more massive than Earth, so its gravity is much stronger than that to which we are accustomed. In order to escape from Jupiter's gravitational influence, you would have to travel faster than 133,018 miles per hour (214,200 kilometer per hour).[1]

A black hole is even more massive than Jupiter. In fact, a black hole is so massive, its gravitational influence is stronger than any other object in the universe. The gravity of a black hole pulls on things so strongly, anything trying to leave its surface would have to travel faster than 186,000 miles per *second* (300,000 kilometers per *second*) to escape. This poses a problem since 186,000 miles per second is the speed of light. Nothing can travel faster than light. Therefore, everything within a black hole—including light—will remain inside, unable to obtain enough speed to overcome its gravitational pull.

Scientists believe that a black hole consists of two parts: a singularity—the center of the black hole—and an event horizon—the boundary marking the beginning of a black hole.[2]

What is a singularity?
The actual black hole, the point where everything ends up.

A singularity could be called the center of a black hole, the point where everything that enters a black hole will eventually end up. The gravity at this point is so strong, everything entering the black hole will collapse together into this infinitesimally small point. As a result, a singularity is incredibly massive, yet takes up no space (volume). Even if more mass is added to the black hole, the singularity will not grow in size. Gravity will continue to squeeze everything down into this inconceivably small point. The concept of a singularity is so difficult to comprehend that even the current laws of physics can't explain it.[3]

What is an event horizon?
A theoretical boundary that surrounds a singularity.
An event horizon could be considered
the "surface" of the black hole.

Event Horizon

Singularity
.

An event horizon marks the boundary of a black hole.[4] Once something crosses this boundary, it must be able to travel faster than the speed of light to escape. Since nothing can travel that fast, everything that crosses the event horizon will remain trapped inside the black hole.

Even though physicist Stephen Hawking theorizes that black holes appear to "radiate" energy, this so-called Hawking radiation (predicted using quantum mechanics, the Heisenberg uncertainty principle, and a bunch of really complicated mathematics) actually comes from an area just outside the event horizon.[5] However, Hawking also predicts that this same effect causes mass to "disappear" from within a black hole (a result of antimatter canceling out matter).[6] Welcome to the bizarre world of quantum mechanics and black holes. The existence of Hawking radiation has been verified mathematically, but physical evidence of this effect has proven elusive.[7]

Once inside the event horizon, everything will eventually end up within the singularity. While the singularity will not grow in size as more mass is added (remember it doesn't take up any volume in the first place), its increasing gravitational strength will cause the event horizon to expand. The black hole that lies in the center of our own Milky Way galaxy is believed to contain the mass of 2.6 million Suns, yet it has an event horizon that would easily fit within our inner solar system.[8]

How do black holes form?
Sometimes with the death of a supermassive star
or sometimes during the formation of a galaxy.
In addition, mini black holes may have formed immediately following the big bang.

No matter how the process starts, the end result is the same. Gravity takes over and collapses everything within reach into an infinitesimally small singularity.

After a supernova has destroyed a large star, the gravity of the remaining stellar debris can be strong enough to cause it to collapse in on itself. If there is enough material, the gravity will be intense enough to collapse even the densest neutron stars. Once the gravity has crushed the sphere of solid neutrons that make up a neutron star, there is no force strong enough to stop the collapse. Gravity continues to compress the debris together until everything that was left over from a star has been squeezed into a singularity.

The size of the black hole depends on how much mass is inside. The greater the mass, the larger the black hole. Large, supermassive black holes are found in the centers of galaxies. Instead of containing all the debris from one star, these black holes contain the material from many different stars. These supermassive black holes are created sometime during the formation of a large galaxy, when stars crowded together within the central core of the galaxy get too close to each other. Gravity attracts these stars to each other. As the stars collide, the gravity of the collective group grows stronger. Eventually, too many stars come together and the overwhelming gravity causes them to collapse in upon themselves. Voilà, an enormous black hole is formed. Astronomers believe supermassive black holes exist in the center of many galaxies, including our own Milky Way and the Sombrero galaxy (as seen below).

On the other end of the size scale, Stephen Hawking has theorized the existence of extremely small, primordial black holes, mini black holes that contain only a thousand million tons of mass (mass equal to that of a large mountain).[9] (While that may not sound small, it is a trivial amount when compared to the mass of a star.) Hawking's calculations suggest that these black holes were formed in the first moment after the big bang (the event that formed our known universe). During the first moment of existence, the pressure and temper-

The Sombrero galaxy (or M104) is believed to contain a supermassive black hole in its core, a black hole large enough to hold one billion solar masses.
Image credit: Hubble Heritage Team (NASA/AURA/STScI)

ature in our universe were so great that it may have been possible for small pockets of matter to be compressed to the point where gravity took over and formed mini black holes. To date, none of these primordial black holes have been observed.[10]

You may wonder how we can talk about various sizes of black holes if the mass entering a black hole ends up inside a singularity that doesn't take up any space. The size of a black hole refers specifically to the size of its event horizon, or surface. As more mass is added to the black hole, its gravity increases. The stronger the gravity, the farther away its effects can be felt—which means a larger event horizon.

If this sounds confusing, don't worry. Even the laws of physics break down at this point. Physics can explain all of the stages involved in the formation of a black hole. However, once a black hole has been created, the physical laws that we understand can't explain the existence of an object of incredible mass yet no volume. We still have much to learn.

How black is a black hole?
Blacker than anything you can possibly imagine.

Even with the use of direct light sources such as the Sun, lightbulbs, candles, and so on, most of what our eyes detect is by way of reflected light. One star (the Sun) can illuminate half our planet because a little of its light is reflected off of everything it shines on. One lightbulb can illuminate an entire room because each surface in your house reflects certain amounts of light. While some surfaces reflect light better than others, almost everything reflects some amount of light for us to see. Everything, that is, except a black hole.

A black hole does not provide a direct source of light by which to see it. Nor does it allow any light to be reflected from it. If, for example, we could aim a huge cosmic flashlight at a black hole, the light beam would simply disappear as it crossed over the event horizon. Absolutely nothing would be reflected back for us to view. Absorbing every particle of light that comes its way would make a black hole blacker than the blackest black you could imagine.

Since black holes are "black," how do you find them?
Astronomers can't see black holes, but using special telescopes they can detect x-rays coming from material about to be sucked into a black hole.

Since nothing can escape a black hole, there is no way we can find one directly. We see stars because of the light they emit. We see a nebula because of the starlight of nearby stars reflecting off the dust and gas. Black holes trap light. They trap everything inside the event horizon, which makes them impossible to see.

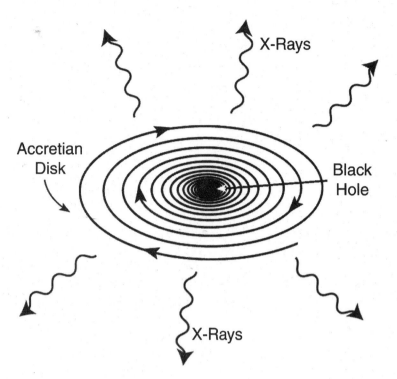

However, a black hole can't swallow everything at once. When material is caught up by the gravity of the black hole, it has to wait in line to pass through the event horizon. This material orbits the black hole, spiraling in toward the event horizon faster and faster, like water being pulled down the drain of a bathtub. This material surrounding a black hole forms into a flattened spiral called an accretion disk. As material spirals in toward the black hole, it heats up to tremendous temperatures and emits x-rays.[11]

Since these x-rays are emitted outside of the event horizon, they can escape. Astronomers can detect these x-rays. If they detect a strong x-ray source and can't see a visible object emitting those x-rays, they may be "seeing" an accretion disk around a black hole. Astronomers have observed several x-ray sources that indicate possible black holes.

What would happen if our Sun became a black hole?
This is not something we have to worry about because our Sun is not nearly massive enough to become a black hole.

Our Sun does not have the mass or gravity necessary for it to collapse into a black hole. So we really don't have to worry about getting sucked into one of these gravitational monsters. However, for the sake of argument, let's imagine that our Sun, by some unfortunate quirk of fate, was replaced with a black hole of equal solar mass. We still wouldn't have to worry—about getting sucked into a black hole, that is. We would have other problems, like lack of heat and light, but our planet would still find itself in its usual orbit, as if nothing traumatic had happened.

The tricky part here is mass. As long as the mass of the black hole equals the mass of the Sun, everything else in our solar system would continue to orbit as if nothing had happened. The reason is based on gravity. The planets are held in their orbits by the gravity of the Sun. The strength of the Sun's gravity is determined by its mass (the amount of matter inside the Sun). If the mass stays the same, the gravity will stay the same, even if that mass shrinks in size.

While there are definitely strong gravitational effects associated with a black hole, they

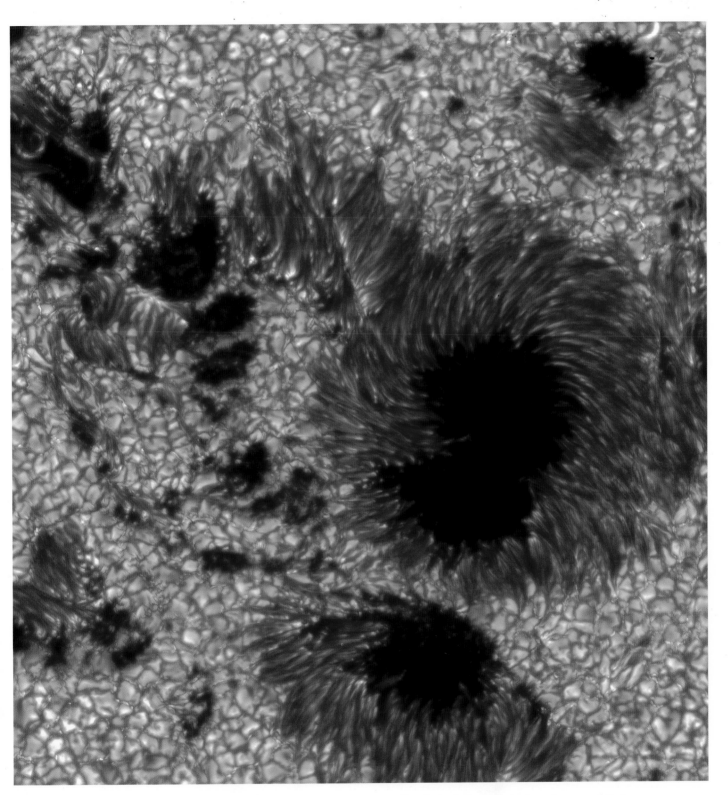

A close-up view of a group of sunspots. Image credit: Royal Swedish Academy of Sciences

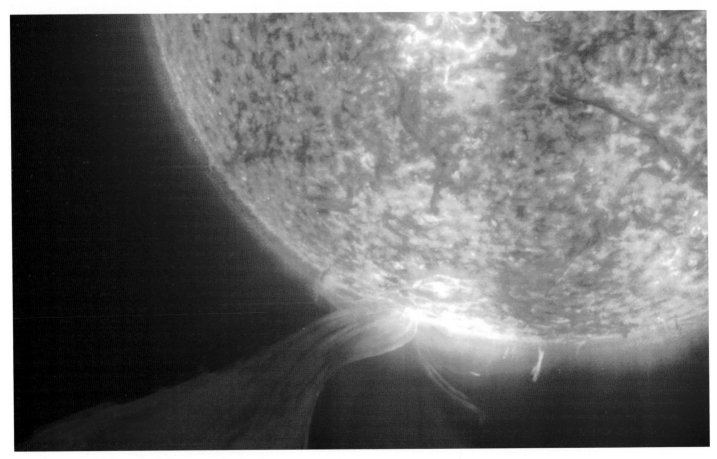
An erupting prominence on our Sun. Image credit: SOHO (ESA & NASA)

A stranger on an alien world, the Mars Exploration Rover *Spirit* took this image of its lander lying on the rusty Martian ground. Image credit: NASA/JPL/Cornell

Look carefully at the tail of comet C2002C1 Linear. You can see hints of both the dust tail and the ion tail. Image credit: Jack Newton, Arizona Sky Village

Io, one of Jupiter's many moons, floats in front of its giant planet. For reference, Io is just slightly larger than our own Moon and lies about twice as far from its planet as our Moon does from Earth. Image credit: NASA/JPL/University of Arizona

The Great Red Spot has been storming on Jupiter for over 340 years. Image credit: NASA/JPL

As the *Cassini* spacecraft approached Saturn, it took this stunning picture of the planet and its famous rings. Image credit: NASA/JPL/Space Science Institute

This false color image taken by the *Cassini* spacecraft allows scientists to detect subtle differences in the composition of Saturn's rings. Image credit: NASA/JPL/University of Colorado

Space Shuttle *Endeavor* approaching the International Space Station. Cook Strait, New Zealand, can be seen in the background. Image credit: NASA

An astronaut works on the Port One truss of the International Space Station. Image credit: NASA

The Eagle Nebula (also known as M 16) is a star-forming region in the constellation of Serpens. This image was taken from an Earth-based telescope. Image credit: Jack Newton, Arizona Sky Village

An outburst from the red supergiant star V838 Monocerotis reveals gas and dust that were ejected from the star at an earlier date. Image credit: Hubble Heritage Team (NASA/AURA/STScI)

The central region of the Eagle Nebula, as seen by the Hubble Space Telescope. Image credit: Jeff Hester and Paul Scowen (Arizona State University) and NASA

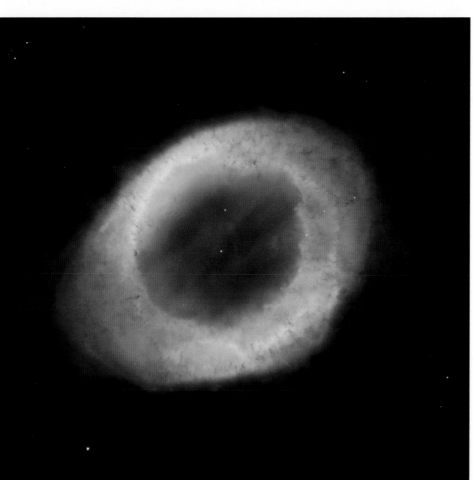

The Ring Nebula, a planetary nebula located in the constellation of Taurus, is an example of what will happen to our Sun when it runs out of fuel. Image credit: Hubble Heritage Team (NASA/AURA/STScI)

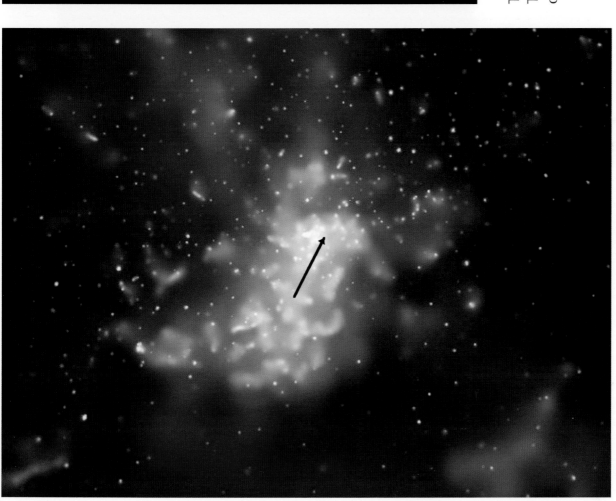

An x-ray view of the center of our galaxy. The arrow points to Sagittarius A*, which is believed to be a supermassive black hole. Image credit: NASA/CXC/MIT/F. K. Baganoff et al.

Galaxies of all kinds can be found in this Hubble Space Telescope Ultra Deep Field Image, an image that required an exposure time of 11.3 days. Image credit: NASA/ESA/S. Beckwith (STScI) and the HUDF Team

occur only within a few hundred miles of its event horizon.[12] A few hundred miles isn't really that far away, especially when you are dealing with the huge distances of outer space. Mercury, the closest planet to the Sun, would be over thirty-six million miles (fifty-eight million kilometers) from this hypothetical black hole.

Could a human travel through a black hole?
Not a chance! You would be toast long before you got close!

Remember, humans can't even travel to the next planet yet—but if you could magically travel to a black hole, you would be killed by many different things before you made it inside.

Within a few hundred miles of a black hole, you would be ripped apart by its strong gravity. The x-rays from the accretion disk would fry you, and the heat would melt even the toughest metals. After you had been ripped apart, fried, and melted, what remained would be sucked into the black hole forever.

Can we send a spaceship through a black hole?
No.

Once again, anything trying to travel through a black hole would be completely and utterly destroyed before it even got to the black hole's event horizon. However, there is really no need to worry about the dangers of black hole travel, especially given the fact that the nearest possible black hole to Earth is several thousand light-years away. With our current technology, it would take us several thousand years to travel just one light-year. We have a long way to go before we have to worry about such possibilities.

Where does everything go after it enters a black hole?
It stays in the black hole.

Everything that enters a black hole remains inside the black hole. Even though everything disappears from our view as it enters the event horizon, astronomers know that everything must still be inside the black hole. How can they be sure the stuff is still there? If it weren't there, there wouldn't be a black hole. If everything that was swallowed by a black hole passed through into another dimension, another time, or another part of space, the black hole soon wouldn't have any mass left. As it lost mass, its gravity would weaken. Without a strong gravity, things would begin to escape and it would no longer be a black hole.

Stephen Hawking has theorized that black holes are leaking mass back into space through a complicated quantum mechanical process that does not violate the speed of light barrier or allow matter to escape from the event horizon (as discussed on page 157). It is Hawking's belief that all matter sucked into a black hole will eventually be returned to normal space.[13] However, this process will take an incredibly long time. For a black hole created after the death of a supermassive star, Hawking predicts it would take 1,000,000,000,000,000,000,-000,000,000,000,000,000,000,000,000,000,000,000,000,000,000 years to evaporate.[14] For comparison, our universe is believed to be approximately 14,000,000,000 years old. In other words, if something of yours gets sucked into a black hole, don't wait around expecting to see it any time soon. And if it does reappear, it certainly won't be in the same shape as it was before it went in!

The fact that the laws of physics can't explain an incredibly massive object that doesn't take up any space makes black holes a fascinating topic. There are many things in the universe that we don't understand completely. Black holes are just one of them. Scientists know enough about black holes to realize we may never completely understand these celestial garbage disposals.

Will black holes devour the universe?
No.

First of all, black holes are rare. For every one galactic black hole, there are billions of stars. Of those billions of stars, only a very few have enough mass to go supernova. Also, not all stars that go supernova will have enough mass left over to form a black hole.

Second, the destructive effects of a black hole only influence things close to it—anything within a few hundred miles of a stellar-type black hole, or within a few million miles of a supermassive black hole. In astronomical terms, a few hundred miles is nothing at all. Mercury is considered to be close to the Sun and it lies at a distance of almost thirty-six million miles (fifty-eight million kilometers). The next nearest star to us is over twenty-five trillion miles (forty trillion kilometers) away.

So the influence of a black hole can only be felt if you are almost on top of it. That, along with the fact that black holes are very rare, makes the universe a very safe place for those of you worried about being swallowed by a black hole.

ABOUT GALAXIES

What is a galaxy?

A huge grouping of stars, planets, star clusters, nebulae, black holes, dust, dark matter, and everything else astronomical.

NGC 4414 is an example of a spiral galaxy. Notice its pinwheel appearance. We are seeing this galaxy almost face-on.
Image credit: Hubble Heritage Team (NASA/AURA/STScI)

Just as a house contains a living room, dining room, bedroom, bathroom, and so on, a galaxy contains stars, planets, moons, nebulae, black holes, and much more. Although it contains many individual items, a galaxy is considered to be the largest astronomical body within our universe. Galaxies themselves vary in size, yet even the smallest one is much larger than anything discussed so far in this book.

What is the Milky Way?

Our home galaxy.

Our solar system is just a small part of the Milky Way galaxy, a spiral galaxy that contains several hundred billion stars. (It's hard to be sure of an exact number, since most of the stars are hidden by large clouds of dust.)

A spiral galaxy is a flat circular disk with a huge sphere of stars in its center. When you look at a spiral galaxy head-on, it looks like a pinwheel. It gets its pinwheel appearance from spiral arms that spread outward from the central bulge, curving through the disk. These spiral arms represent regions of active star formation. Although the arms are very distinct when viewed head-on, they disappear when viewing the galaxy from its side. Seen edge-on, a spiral galaxy's disk is very thin while its central bulge forms a bright ball in its center. Globular clusters can be found both above and below the galaxy in a region called the halo.

The disk of our Milky Way galaxy is believed to be approximately 160,000 light-years in diameter, but only about 2,000 light-years thick. The bright sphere of stars in its center is somewhere between 7,000 and 10,000 light-years in diameter. Astronomers have determined that our solar system lies in a spiral arm approximately 26,000 light-years from the galactic core.[1]

If we could view our Milky Way from the side, it would look much like this galaxy, NGC 891. Image credit: Jack Newton, Arizona Sky Village

When can I see the Milky Way?
Any time you look up (or down)!

During the day, with the bright Sun in the sky, you are seeing a star that belongs to our Milky Way galaxy. If you go outside on any clear night and look up, all of the natural objects you will see (with the exception of three fuzzy blobs) belong to our Milky Way galaxy. Remember, our solar system is part of the Milky Way. As such, the Moon and planets can be counted as members of our galaxy. (You will probably also see human-made airplanes and satellites in the night sky. These don't really count.) The stars are our more distant galactic neighbors.

In addition to all of the individual stars, the main portion of our galaxy can be seen as a hazy streak of light in the evening skies from late August through October. During this time of year, the nighttime side of Earth is looking toward the central bulge of our galaxy. The "milky" streak we see is the disk of our galaxy extending out from opposite sides of the bulge.

The Andromeda galaxy has two small elliptical galaxies for companions.
Image credit: Jack Newton, Arizona Sky Village

The bulge itself, located in the constellation of Sagittarius, is difficult to see. Although this region contains an enormous number of bright stars in a tight, compact ball, large amounts of interstellar dust lie between us and the galactic core. These dust clouds block our view.

The milky streak we can see is made up of the faint, more distant stars within the Milky Way's disk. These countless stars are too far away to be seen as pinpoints of light. They all blur together to form the hazy cloud. However, even with a small pair of binoculars, you can begin to make out some of the thousands of stars.

As mentioned at the beginning of this answer, there are three objects you can see without a telescope that don't belong to our galaxy. In the Northern Hemisphere, one object is the Andromeda galaxy. This faint, fuzzy blob, located in the constellation of Andromeda, is an entirely separate galaxy that contains its own stars, star clusters, nebulae, black holes, and more. The Andromeda galaxy is 2.2 million light-years away from us and is the most distant object you can see without the aid of binoculars or a telescope.[2]

In the Southern Hemisphere, you can see two galaxies without a telescope, the Large and Small Magellanic Clouds. These two galaxies are companions to our own Milky Way. The Large Magellanic Cloud is about 170,000 light-years away, while the smaller galaxy is about 200,000 light-years from us.[3]

From Earth, why doesn't the Milky Way look like a spiral galaxy?
Because we are inside it, looking out.

For the same reason that you can't see the forest for the trees, we can't see our spiral galaxy because of all the nearby stars and dust clouds. Also, at any given time, we are viewing only a small part of the stars that make up the galaxy—and only the brightest stars at that. You must also take into account the large quantities of dust that block our view of many of the more distant stars. All these factors work against providing us with a full view of the beautiful spiral arms of our own galaxy.

Despite the difficulties presented by our location deep within the Milky Way, astronomers have been able to put together a map of our galaxy that shows an edge-on view of its disk and central bulge. The most interesting of these maps was constructed from data collected by satellites designed to detect infrared radiation. This type of radiation is made up of longer wavelengths than visible light. It is more difficult for interstellar dust particles to absorb or scatter these longer wavelengths. Consequently, infrared radiation allows astronomers to see through much of the interstellar dust that lies between us and the center of our galaxy. The infrared maps reveal a shape almost identical to other edge-on spiral galaxies that we see at different locations throughout the sky.

Do we have a picture of our galaxy?
We have pictures of parts of our galaxy, but we don't have a picture showing the entire Milky Way galaxy from a distance.

Imagine trying to take a picture of your house. The first thing you would have to do is get outside of the house. Then, it is doubtful you could simply stand on your doorstep and take the picture. In order to include the entire house in your camera's field of view, you would probably have to walk away from the house for a bit. Just how far away you would need to walk depends on how large the house is. For even the biggest house, this distance shouldn't be too far. Distance, however, is definitely a problem when trying to take a picture of our entire galaxy.

We live inside the Milky Way galaxy. To photograph the galaxy as a whole, we would have to travel thousands of light-years away from it before we could get the entire galaxy into a camera's field of view. Recall that one light-year is 5.9 trillion miles (9.4 trillion kilometers). Our galaxy is 160,000 light-years in diameter and about 2,000 light-years thick. That would be some trip just for one picture!

The farthest we have sent a robotic spacecraft is just beyond the orbit of Pluto, over 8.8 billion miles (14.3 billion kilometers) away. Right now, human beings can't even travel a few million miles to the next planet. So a spectacular picture showing the spiral arms of our Milky Way galaxy is probably a very long time coming.

What's it like at the center of our galaxy?
Because a supermassive black hole lies at the core of our Milky Way, conditions there would be extremely harsh.

Within the central bulge of our galaxy lies a whirling maelstrom of stars, hot gases, strong magnetic fields, and intense gamma ray, x-ray, and infrared radiation sources. If that doesn't sound threatening enough, you would also find, at the very center, a supermassive black hole.

Dubbed Sagittarius A* (the * indicates that this object is a point source instead of a cloud or cluster of stars), this black hole is believed to contain the mass of 2.6 million Suns.[4] Even with such an enormous mass, its event horizon is so small that it could easily fit within our inner solar system. While astronomers are unable to see this black hole directly, they have detected flares of extremely strong x-rays coming from material they believe to be trapped in the black hole's accretion disk. In addition, they have detected nearby stars traveling in orbits with velocities as high as hundreds of miles per second. Only a star caught in the gravitational pull of an extremely massive object (a black hole) would be able to travel at such high velocities.[5] (For more information on black holes, see "About Black Holes" beginning on page 155.)

An x-ray image of the center of our galaxy can be found in the color insert section of this book.

So you would find the center of our galaxy a very unpleasant destination, one you wouldn't be able to write home about because its intense gravity, radiation, and heat would annihilate you long before you got close.

Are there different types of galaxies?
Yes.

Though spiral galaxies may be the most magnificent to look at, they are not the only galaxies in the universe. There are elliptical galaxies, irregular galaxies, and peculiar galaxies as well.

A small portion of the Hubble Space Telescope's *Ultra Deep Field image showing many different types of galaxies can be found in the color insert section of this book.*

Elliptical galaxies are huge balls of stars with no spiral arms. They range in shape from an almost-perfect sphere to an extreme oval. The largest and smallest known galaxies are both of the elliptical variety. Irregular galaxies don't have any particular pattern to them. The Large and Small Magellanic Clouds, two of the Milky Way's companion galaxies, are both irregular galaxies.

Peculiar galaxies are just that—peculiar. Something strange is happening in these galaxies and astronomers aren't certain what that is. A peculiar galaxy may have a spiral or elliptical shape, but it will also have huge amounts of energy streaming out of it. Where this energy is coming from and what is sustaining it is something astronomers are still trying to figure out.

What is the closest galaxy to the Milky Way?
Canis Major Dwarf

This small galaxy lies just 25,000 light-years from our solar system and 42,000 light-years from the Milky Way's galactic core.[6] Canis Major Dwarf contains only about one billion stars, a very small number for a galaxy. How this galaxy ended up with so few stars may be explained by its close proximity to the Milky Way. Gravity from our own much larger galaxy appears to be stealing stars from this small companion. It may be that in a few billion years, our galaxy could completely gobble up this much smaller star group.

Even though Canis Major Dwarf is so close to us, large clouds of interstellar dust within the Milky Way are blocking our view of this small galaxy, making it extremely difficult to see. That is why Canis Major Dwarf was not discovered until November 2003 by astronomers doing an infrared survey of the sky. It may be possible that other small companion galaxies exist that have yet to be discovered.

What is the most distant galaxy?
Abell 1835 IR1916

To date, the record for the most distant galaxy belongs to an extremely faint fuzzy blob known as Abell 1835 IR1916. Astronomers calculate that this galaxy lies at an astonishing distance of 13.23 billion light-years.[7] In other words, light that we are seeing from this galaxy left the star group 13.23 billion years ago. If you consider that the universe was formed 13.7 billion years ago (as many astronomers believe), this means that Abell 1835 IR1916 was formed a mere 470 million years after the big bang.[8] This makes it the youngest star group astronomers have ever observed. (While 470 million years ago may seem like a long time by human standards, keep in mind we are talking about star time. Anything under a billion years is more like a blink of an eye.)

The record for the most distant galaxy has changed frequently in the last few years as technological advances have enabled astronomers to take photographs of fainter and fainter objects. The current distance record holder, Abell 1835 IR1916, was announced in March 2004, beating out another galaxy (near Abell 2218) whose distance of 13 billion light-years was announced the month before.[9] How far away will astronomers be able to see? Will we ever see the remnants of the big bang? Stay tuned.

How far apart are galaxies?
Generally several thousand light-years apart.

In this ever-changing universe, everything is in a constant state of motion, even giant galaxies. Therefore, the average distance between galaxies can vary greatly. Some galaxies are moreover colliding with each other, while others are moving away at tremendous velocities.

In the Local Group of galaxies, of which our Milky Way belongs, there are almost forty galaxies within a three-million-light-year block of space. Within this group, the closest galaxy to us, Canis Major Dwarf, is a mere 25,000 light-years away, while the most distant, the Andromeda galaxy, is about 2.2 million light-years. In addition to the Local Group, there are clusters and super clusters of galaxies. So when determining how far apart galaxies are, it really depends on where you are looking.

How many galaxies are there?
More than we can count.

There are more galaxies in our universe than there are stars in our sky. There are large catalogs filled with exact coordinates for thousands of known galaxies that can be seen with Earth-based telescopes. In addition, astronomers aimed the Hubble Space Telescope at an "empty" part of the sky and took an eleven-day exposure to see what they could find. The resulting image revealed an amazing sight—thousands of distant galaxies, much too faint to be seen by normal means. And that was just a very small section of "empty" space. In other words, there are more galaxies in our universe than we could ever count.

See just a small portion of this Ultra Deep Field image in the color insert section of this book.

ABOUT TELESCOPES

What is a telescope?
Who invented the telescope?
What determines the size of a telescope?
Are there different types of telescopes?
What type of telescope is the best?
What should I look for when buying a telescope?
What will I be able to see through a telescope?
Will the object I see through a telescope look as nice as its picture?
What is the largest telescope?
How big is the Hubble Space Telescope?

What is a telescope?
An instrument that uses a combination of lenses and/or mirrors to magnify our view of distant objects.

All telescopes, large or small, cheap or expensive, are built with the same goal in mind—to gather as much light as possible from a target object, then magnify that light by spreading it out over a larger and larger area. Notice that the first priority of a telescope is to gather light —not to magnify. Despite all the advertising, magnification comes in a distant second in the list of telescope priorities.

Every telescope (no matter what model or size) will have the following common components: a primary lens or mirror, an optical tube, and an eyepiece. The primary lens or mirror is used to collect light from the desired target. That light travels down the optical tube, through the eyepiece, and into your eye. Depending on the model, size, and gadgetry, the above description may change slightly, but not by much.

Who invented the telescope?
A Dutch spectacle maker by the name of Hans Lippershey invented the telescope in 1608.[1] However, the first person to turn a telescope to the sky was the Italian physicist Galileo Galilei.

In 1610, using a small refracting telescope of his own construction, Galileo was the first to see craters on the Moon. He was the first to note that Venus goes through phases like our Moon and that Saturn had two small bumps, one on either side of the planet. (Through a better-quality telescope, these "ears" [as Galileo called them] turned out to be Saturn's beautiful ring system.) He was also the first to see the four largest moons of Jupiter—a discovery that helped to get him in trouble with the Catholic Church.[2]

As Galileo observed Jupiter through his telescope, he could see four starlike objects near the planet. From night to night, these four objects changed positions, but never strayed far from Jupiter. Sometimes, one or more of them would disappear, only to reappear a few nights later. The more he observed, the more Galileo realized he was seeing four moons in orbit around Jupiter. This discovery went against the Church's belief that Earth was the center of the universe.[3] Everything was supposed to revolve around our own planet.

In 1633, after a long battle of wits, Galileo was forced by the Inquisition to denounce his findings and accept the idea of an Earth-centered universe. In addition, the Church placed him under house arrest where he remained until he died in 1642.[4] In 1992, three hundred and fifty years after his death, the Catholic Church announced that it had come to terms with Galileo's discoveries.[5] A little late for Galileo, a man who was simply telling people what he saw through his small telescope.

What determines the size of a telescope?
The diameter of its primary mirror or lens.

Most people think a telescope's size is based on the length of its optical tube. This may seem like a natural assumption, but as is the case with many assumptions, it's not correct. The problem with a telescope's length is that it does not provide a reliable measurement of the instrument's performance. However, the size of a telescope's primary mirror or lens does. For example, it is possible for a 2-inch (60 mm) telescope to be anywhere from one to three feet in length, depending on the curvature of its primary lens. Viewing the same object through two 2-inch telescopes of different lengths will show you almost the same thing. Viewing the same object through a 3-inch (90 mm) telescope will show a brighter, sharper object, no matter what the telescope's length. Why is that? The 3-inch primary lens collects more light than the 2-inch lens and the resulting image is therefore much brighter.

Because telescopes with primary lenses and mirrors of the same size are comparable in performance, astronomers decided that this would be a good way to define a telescope's size. Still, whether you use English units (2-inch) or metric units (60 mm) to identify your telescope, that is up to you.

Are there different types of telescopes?
Yes. There are three basic types of telescopes: a refractor, a reflector, and a catadioptric telescope.

While all telescopes are built with the same goal in mind—to gather as much light as possible—their designs can vary greatly. A refractor is the most basic telescope and has been around the longest. (Galileo's telescope was a refractor.) It uses a primary lens to bring light to a focus. A reflector uses a primary mirror instead of a lens, while a catadioptric telescope uses both a lens and a mirror to do the job.

While there are many variations within these three categories, the following diagrams show the basic structure of each telescope type.

A refractor uses a lens to collect light.

Small refractors, 2–3 inches (60–90 mm) in size, are the cheapest and most common type of telescope available. They can be purchased from any number of places, including retail stores, camera shops, malls, and online. Larger refractors, 4–7 inches (101–178 mm) in size, do exist, but they are rather expensive. The higher cost is directly proportional to the larger lenses. Because light has to travel all the way through the primary lens, the glass used in these lenses must be of the highest quality. The larger the lens, the more difficult it is to get a high-quality, perfectly shaped piece of glass.

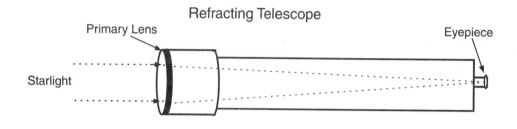

A reflector uses a mirror to collect light.

Typical reflecting telescopes come in sizes ranging from 4.5 to 10 inches (114 to 254 mm). The limiting factor with these telescopes is the length of the optical tube. As a reflector's primary mirror increases in size, so does the distance light must travel through the telescope before it can come to a focus. This distance is called the focal length. Longer focal lengths make for longer and more awkward optical tubes.

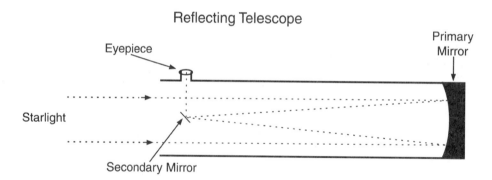

There are two basic types of reflectors, Newtonian and Dobsonian. The main difference between the two is the type of mount that is used to hold the telescope in place. (See "Considering the Mount" on page 180 for a description of different telescope mounts.) Newtonian telescopes use an equatorial mount. Common sizes for Newtonian telescopes are 4.5–6 inches (114–152 mm). Dobsonian telescopes have a very basic Alt/Az mount that allows them to hold a much larger optical tube. Common sizes for Dobsonian telescopes range from 6 to 10 inches (152 to 254 mm), although the author has seen much larger ones, including a homemade 40-inch (1,016 mm) telescope.

A catadioptric telescope uses a combination of lenses and mirrors to collect light.

Within this category, there are three common types: a Maksutov-Cassegrain telescope, a Schmidt-Newtonian telescope, and a Schmidt-Cassegrain telescope. All use a lens at the front of the telescope to correct for spherical aberation and a primary mirror at the back of the telescope. The lens in this system is referred to as a correction plate.

A Maksutov-Cassegrain telescope uses a correction plate/mirror combination to fold the light several times in a short amount of space. As a result, these telescopes have fairly short optical tubes and are easy to transport. Common sizes for this telescope range from 3 to 7 inches (90 to 178 mm).

Maksutov-Cassegrain Telescope

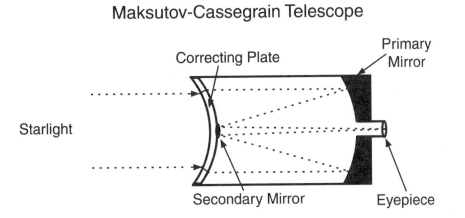

For a Schmidt-Newtonian telescope, a correction plate has been added to the front of a basic Newtonian reflector allowing for larger mirrors to be used without adding additional length to the optical tube. Common sizes for this telescope range from 6 to 10 inches (152 to 254 mm). Schmidt-Newtonian telescopes are usually the least expensive of the catadioptric telescopes.

Schmidt-Newtonian Telescope

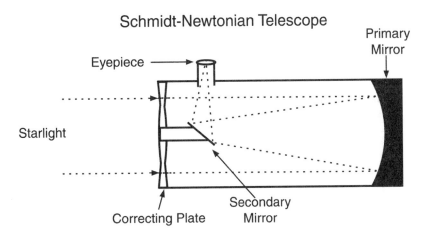

A Schmidt-Cassegrain telescope also uses a correction plate/mirror combination to fold the light several times in a short amount of space. This particular combination allows for the use of larger primary mirrors. Common sizes for this telescope range from 8 to 16 inches (203 to 406 mm).

Schmidt-Cassegrain Telescope

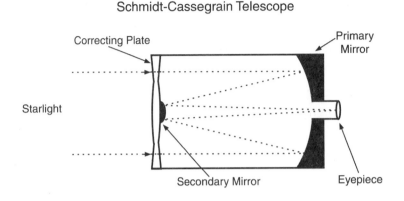

What type of telescope is the best?
There is no easy answer to this question. It depends on what you want out of your telescope and how much you are willing to spend.

There are trade-offs between each type of telescope. A refractor is usually the easiest telescope to operate. However, the most common refractors can also be the most frustrating ones to use. Their small primary lenses don't gather much light, and as a result, all but the Moon and brightest planets are rather faint and hard to see. The larger reflecting telescopes offer a bigger primary mirror, but in a more complicated layout. The catadioptric telescopes offer bigger primary mirrors in an easier to carry package. However, since these telescopes use both mirrors and lenses in their systems, they are more expensive. If money is not an issue and you really want to see great views of the Moon and planets, a large refractor is the telescope for you.

No matter what you are willing to spend, and what size of telescope you buy, it will take practice and patience to get the most out of your eye-to-the-sky.

What should I look for when buying a telescope?
There are a variety of items to consider when purchasing a telescope, including cost, size, magnification/power, tripod, mount, portability, and accessories.

Trying to sort through all the information involved with purchasing a telescope can be more intimidating than the vast reaches of space you hope to explore. Below are some items to consider before you hit the stores or Internet.

Considering Cost: With a telescope (as with most things in life) you get what you pay for. If you want a good astronomical telescope—one designed to deliver gorgeous views of the heavens—you will have to spend more than $100.00. Below are some cost guidelines.

- **Less than $100.00:** For an astronomical telescope, don't waste your money. The quality of the optics and the construction of the telescope and mount will probably lead to more frustration than it is worth.

- **$100–$300:** Ask some questions, shop around, and you can get a decent beginning telescope in this price range. Telescopes in this price range are usually small refracting telescopes between 2 and 2.5 inches (60 and 70 mm) and the occasional 4.5-inch (114 mm) Newtonian reflector telescope.

- **$300–$500:** In this price range, you should be able to find high-quality beginning and even some good-quality intermediate telescopes. Telescopes in this price range include: 2–3-inch (60–90 mm) refractors and 4.5–6-inch (114–152 mm) reflectors.

- **$500+:** As in any hobby, you can spend as much money as you want. In this case, more money means that you will be able to afford a telescope with a fairly large primary mirror or lens. Telescopes in this price range include larger refractors and catadioptric telescopes.

Considering Size: A telescope is essentially a light bucket. You want to gather as much light as you can, so you want the biggest "bucket" you can afford. Remember that the size of a telescope depends on the diameter of the largest mirror or lens in the system—not the length of the telescope tube. Typical telescopes range in size from 2 to 16 inches (60 to 406 mm).

Considering Magnification or Power: Magnification, or power, *is not* an important item to consider when purchasing an astronomical telescope. (For clarification: advertisers use the words magnification and power to describe the same thing.) For smaller telescopes, the average range of useful magnification is anywhere from 40x to 150x. Higher magnification will make objects larger, but much fainter and more difficult to see.

When you magnify something, or make something appear larger, you are taking the light collected by the telescope and spreading it out over a larger and larger area. If you are looking at a bright object—such as the Moon—there is plenty of light to spread around, so more magnification is a good thing. If you are looking at faint objects—like most everything else in the night sky—high magnification doesn't work very well. Using high magnification on an extremely faint object merely spreads the already faint light over a greater area, thus making it even harder to see. So forget about magnification and worry about the size of the telescope's primary lens or mirror!

Considering the Tripod: A steady tripod is an absolute must for an astronomical telescope. Your tripod should be heavy-duty and able to withstand slight bumps or the occasional breeze without moving the telescope that it's holding. You can test a telescope's tripod while it is still in the store. Aim the telescope at an object across the store then gently bump the tripod.

See how much force it takes to knock the telescope off of its target. You want a tripod to hold the telescope securely so that when you are fumbling around in the dark, the telescope will remain in one place.

Considering the Mount: A mount refers to the way a telescope is attached to the tripod. There are two basic types of mounts: alt/az and equatorial.

Alt/Az Mount: Short for altitude/azimuth, the name refers to how the telescope is adjusted while on this mount. *Altitude* refers to the up and down motion of the telescope while *azimuth* refers to its side to side motion. An alt/az mount is the simplest to set up. It is also the cheaper of the two types of mounts, as well as the lighter. One drawback of this mount is that it requires constant adjustment in both altitude and azimuth to keep an astronomical target in sight.

Equatorial Mount: When an equatorial mount is aligned properly, you will only need to adjust the east/west motion of the telescope to counteract the Earth's rotation. This adjustment can be made manually or with a drive motor. An equatorial mount is more complicated than an alt/az mount. It is thus heavier, more expensive, and more difficult to set up. Once aligned, however, this mount makes it easier for the observer to keep objects in the telescope's field of view.

Considering Portability: If you are like most of the world, you live in a populated, well-lit area. To get a good view of the night sky requires putting the telescope in the car and driving to a dark spot to observe. Because of this fact, you might want to consider how portable the telescope is before you make a purchase. Just a few things to consider . . .

- Will it fit in your car?
- How heavy is it? (Can you lift it by yourself?)
- Does it require power? If so, can it run on batteries?

Considering Accessories: Accessories will enhance your telescope's performance, but you probably don't need every one that is offered. It is really up to the individual as to when and what type of accessories you want or need to purchase. Below are brief descriptions of some typical accessories beginning with the most important but often overlooked accessory—the eyepiece.

- **Eyepieces:** Most telescopes allow the observer to change eyepieces. An eyepiece is identified by its focal length (in millimeters), which can be found stamped on its side. Since magnification is equal to the focal length of the telescope divided by the focal length of the eyepiece, you change the magnification of your telescope every time you change eyepieces. Normally, eyepieces range in size from 4 to 40 mm, with the smaller focal length eyepieces magnifying the most.

The most useful eyepieces for a small telescope are between 12 and 40 mm. When you are starting out, stick with the larger numbered eyepieces. With less magnification, you will be able to see a larger area of sky, making it easier to find specific targets. You can always increase magnification by changing the eyepiece after you have the desired object centered in the telescope.

Barlow Lens: A Barlow lens is placed between the telescope and the eyepiece and doubles or triples the amount of magnification of your eyepiece. With small telescopes, Barlow lenses are not usually recommended. When you magnify an object, you don't increase the amount of light coming in, you only spread out what little light there is. A Barlow lens spreads the light out so much that the image is very faint and difficult to see. It is also difficult to look through and requires some practice before you get the hang of it. It is a good idea to put the Barlow lens aside until you are comfortable with the way your telescope works. Then try using the Barlow lens on bright objects such as the Moon or Jupiter.

Drive Motor: A drive motor attaches to an equatorial mount. When activated on a properly aligned telescope, it will slowly move the telescope just enough to counteract Earth's rotation and keep the telescope pointing at your desired target. For a small telescope, a drive motor is nice to have but not absolutely necessary. With a little practice, you can learn to manually move the telescope to keep your target in sight.

If, however, you are planning to have large groups of people look through your telescope, a drive motor is recommended. With a drive motor in place, you can relax and don't have to check the telescope's position after every other viewer. A drive motor is also recommended for anyone interested in astrophotography. Larger telescopes usually come with a drive motor built into the system.

Elbow, or Right-Angle Viewer: An elbow, or right-angle viewer, is shaped like an "L." An eyepiece fits into one end of the "L." The other end is placed in the telescope. Inside the elbow is a prism or mirror that takes the light from an object and reflects it ninety degrees into the eyepiece. This is very useful when you are trying to look at something straight overhead. An elbow, or right-angle viewer is not necessary, but it really makes viewing more comfortable.

Electronics: Today, many telescopes come with optional electronic aids that offer everything from simple "go to" (GOTO) capabilities to incorporated global positioning systems (GPS). These electronic aids have made it much easier for observers to locate objects in the sky even if they don't know the difference between Orion and Leo. For example, these aids can tell when and where each planet is visible and where different constellations are located in the sky. They can also accurately point your telescope toward any number of faint celestial objects, including star clusters, nebulae, galaxies, and more. While it sounds incredibly easy, there are a few catches. The basic electronic packages will require you to align your telescope properly before beginning. It may also ask you a few questions to determine your location.

These optional electronic packages are changing rapidly as technology improves, so it is difficult to be specific. However, if the telescope you choose has an electronics package you can afford, it is worth having. Instead of spending all night looking for one planet, you will be able to observe as many objects as you like. And if you want to take a crack at the night sky without aids, you can always turn your electronic gizmo off.

☸ **Finderscope:** A finderscope is located on the optical tube of your telescope. It is a small telescope with less magnification and a greater field of view than the primary telescope. A finderscope is held in place with a number of screws that allow you to adjust where it is pointing. Through a finderscope, you can see a large area of sky, which will help you locate the object you want to observe. Centering an object in the finderscope will also center it within your main telescope if—and this is a big if—you have aligned the two telescopes before your observing run. If your telescope comes with a finderscope, it is important to realize that the factory does not align the two telescopes. That job is up to you. Be sure and align the two telescopes *before* your observing run, preferably during the day when you can see what you are doing.

☸ **Moon Filter:** A moon filter is a small disk that screws into an eyepiece and reduces the amount of light coming into your eye. While somewhat useful, it is not really necessary. During those times when the Moon is painfully bright, a moon filter might be nice to have. However, a small piece of cardboard or cloth covering a portion of your telescope's objective lens or mirror will work just as well to block out some of the light, as will a pair of sunglasses. Seriously. Try it before you laugh.

☸ **Slow-Motion Controls:** When observing through a telescope, there will be times when you want to move the telescope quickly (such as moving from one part of the sky to another) and other times when you will want to move the telescope slowly (such as moving from one lunar crater to another). All telescopes have ways to move the telescope quickly. Slow-motion controls (which allow you to move the telescope slowly) are sometimes optional accessories on small telescopes. These controls are extremely useful, especially if your telescope doesn't have a drive motor. With slow-motion controls, it is possible to make the small corrections necessary to counter Earth's rotation and keep an object in view. Even if your telescope has a drive motor, these controls will prove handy. Their use makes the telescope more manageable and your overall observing experience less frustrating.

☸ **Sun Filter:** Some telescopes come with a small sun filter that screws into the eyepiece like a moon filter. If you have one of these filters, THROW IT AWAY! These filters are made of smoked glass and can shatter if they get too hot. If a filter shatters while you are looking at the Sun, the magnified sunlight will burn your eye before you can pull away. You would be blinded in that eye.

The Sun is a dangerous object to observe even without a telescope. So play it safe. Don't look at the Sun unless you are with someone you are certain knows what she is doing *and* has special filters that are placed over the end of the telescope. Even though the Sun is a fascinating object, it is not worth losing your eyesight over. If you're not sure, don't look!

No matter what telescope you end up with or what accessories you buy, please keep this one fact in mind: without a little practice, a perfectly decent telescope will end up in the closet after one failed observing attempt. Start slow and make a few practice runs with your telescope during the day (so you can see what you are doing). It will make all the difference in the world.

What will I be able to see through a telescope?
Craters, ridges, and mountains on the Moon; the rings of Saturn; several moons of Jupiter; a nebula where stars are being formed; and much more—as long as you take the time to look.

Today, with all of the robotic exploration of our solar system, we have grown used to fabulous, close-up images of planets, star clusters, nebulae, and more. These images have set a standard that is much higher than the average telescope can match. As a result, most people are disappointed the first few times they look through a telescope. Keep the faith. The universe is a beautiful place. It just takes patience and practice to bring it into view. Below are a few objects to start with.

- **The Moon:** The Moon is large and bright, making it the easiest object to find in the night sky. Because it is so close to Earth, even the smallest telescopes can show detailed views of its surface, including craters, mountains, and valleys.
- **The Planets:** In the night sky, Venus and Jupiter are the brightest of all the other planets in our solar system, making them the easiest planetary targets for your telescope. Of the two, Jupiter is by far the most interesting. With a small telescope, you can see cloud bands on the planet, the Great Red Spot—if it is facing Earth at the time you are looking—and up to four of Jupiter's many moons. Although Saturn is fainter, a small telescope can show you its beautiful ring system. Astronomy software

packages and monthly astronomy magazines like *Astronomy* and *Sky & Telescope* can help you locate the ever-moving planets.

- **The Stars:** Because stars are so far away from us, they appear as pinpoints of light, no matter how large the telescope. However, they are still intriguing things to look at through a telescope. They are bright, and are good objects with which to practice your observing skills. Some stars even have a hint of color to them.

- **The Orion Nebula:** Located in the sword hanging off of Orion's belt, this diffuse nebula (cloud of gas and dust) is a birthplace of stars. Through a telescope, you can see a wispy cloud of gas surrounding a small group of stars. The Orion Nebula (also known as M42) is best seen in the evening sky during the months of December through March. (Monthly astronomy magazines or astronomy software can help you locate this bright nebula.)

The Orion Nebula. Image credit: Jack Newton, Arizona Sky Village

The more you practice with your telescope, the more you will be able to see. It will take time, but hang in there. The night sky is worth it.

Will the object I see through a telescope look as nice as its picture?
If you are looking at the Moon, yes. If you are looking at any other astronomical object, no.

Since the Moon is so bright and close to Earth, it is an awesome sight to look at through even the smallest telescopes. Everything else in space is too distant and too faint for our eyes to see much detail, even through a really large telescope. Astronomers get around this problem by taking pictures.

The reason photographs of astronomical objects look better than what you can see through a telescope is because our eyes are not as sensitive as film or electronic detectors. To get those beautiful astronomical images, astronomers attach a film or CCD (charge-coupled device) camera to a telescope and take long exposures of their objects. They leave the shutter of their cameras open for anywhere from several seconds to several hours so that the film or electronic chip can collect as much light as possible from the faint target object. When the exposure is complete, the resulting image is a combination of all the light the camera received during the length of the exposure.

The human eye, on the other hand, has an exposure time equal to about one-thirtieth of a second. This means that every thirtieth of a second, the eye sends a new signal to the brain for processing. This rapid refresh rate is great for allowing us to keep track of all the movement around us. It is poor, however, for seeing faint astronomical objects.

That said, there are still many reasons to look at those faint fuzzy blobs through a telescope. First of all, when you look at something through a telescope, it's real. It is right there in front of you. Second, if you practice a bit, you will begin to see subtle differences in those fuzzy blobs. They are not all alike. Soon you will be able to tell a galaxy from a globular star cluster and a planetary nebula, and that is even more satisfying than looking at those pretty pictures someone else took!

What is the largest telescope?
The 10-meter Keck I and Keck II telescopes on Mauna Kea in Hawaii.

Together, the two Keck telescopes hold the current record for the world's largest telescope. They are identical reflecting telescopes that began observing in 1993 (Keck I) and 1996 (Keck II).[6] Instead of one 393-inch (10-meter) primary mirror, each telescope has thirty-six individual mirrors mounted in a honeycomb structure. Each one of these mirrors is hexagon-shaped and 72 inches (1.8 meters) in diameter. Sensors on the back of the mirrors keep them aligned with each other and allow them to act as one single large mirror. Building and maintaining thirty-six small mirrors instead of one enormous mirror made the telescope lighter and easier to handle.

Other large telescopes include: the 363-inch (9.2-meter) Hobby Eberly telescope at McDonald Observatory in Texas; two 331-inch (8.4-meter) telescopes that work together to form the Large Binocular Telescope on Mt. Graham, Arizona; four 323-inch (8.2-meter) telescopes that work together as one to form the Very Large Telescope in Cerro Paranal, Chile; the 323-inch (8.2-meter) Subaru Telescope on Mauna Kea; the 315-inch (8.0-meter) Gemini North telescope on Mauna Kea, Hawaii, and its twin, the 315-inch (8.0-meter) Gemini South telescope in Cerro Pachón, Chile.[7]

As technology advances, so does the size of primary mirrors and the ability to combine light from several different sources. Who knows how large the telescopes of tomorrow will be? Currently there are plans for a 1,181-inch (30-meter) telescope to begin construction in 2012.[8] One wonders what we will see when this telescope comes on line.

How big is the Hubble Space Telescope?
94 inches (2.4 meters) in diameter.[9]

Compared to the telescopes listed in the last question, the Hubble Space Telescope is rather small. However, its location more than makes up for its relatively small diameter. Floating in low Earth orbit, this telescope isn't affected by Earth's turbulent, sometimes hazy, sometimes cloudy atmosphere. Astronomers are guaranteed a crystal-clear view. Hubble has thus provided us with some of the most beautiful and detailed images of the universe ever taken.

PART 2

QUICK FACTS

This section is designed to provide basic information about our solar system, nearby stars, and constellations in an easy-to-access format. Instead of wading through detailed descriptions and explanations, this section offers you just the facts—basic information about the planets, the Sun, the Moon, eclipses, meteor showers, and more.

There are many instances where the provided information will prove useful: for example, if you are planning a camping trip and need to know if there are any meteor showers or eclipses you should watch for, this section contains all you need to know. Or would you like to find a Web site that has some great astronomical images or facts about the space program? This section offers some great Web sites to get you started.

In addition, have you ever wondered when you could see a First Quarter Moon, or what they named all those moons of Jupiter? Have you wondered which are the five brightest stars in the sky, or how many constellations there are? Below is a guide to what you can find in this section, including answers to all the previous questions.

QUICK FACTS ABOUT:

The Sun
The Planets
The Moon
The Phases of the Moon
Moon Totals for Each Planet
Moon Names
The Biggest Moons of the Solar System
Lunar Eclipses for 2005–2020

Solar Eclipses for 2005–2020
Meteor Showers
The Twenty-five Brightest Stars
The Twenty-five Closest Stars
The Stars of the Big Dipper
The Constellations
Astronomy and Space-Related Web Sites

THE SUN

Distance from Earth .... Average: 92,900,000 miles (149,600,000 kilometers)

Maximum: 94,450,000 miles (152,100,000 kilometers)

Minimum: 91,350,000 miles (147,100,000 kilometers)

Diameter .... 864,400 miles (1,392,000 kilometers)

Temperature .... Surface: 10,000°F (6,000°C)

Core: 27,000,000°F (15,000,000°C)

Rotation Rate .... Equator: approximately 25 days

Poles: approximately 35 days

Composition .... 75% Hydrogen

24% Helium

1% other gases

Average Density* .... 1.4 g/cm^3

Mass .... 4.38×10^{30} pounds (1.99×10^{30} kilograms)

Information gathered from the following sources: Roger A. Freedman and William J. Kaufmann III, *Universe*, 6th ed. (New York: W. H. Freeman and Company, 2002), pp. 390–413; Jay M. Pasachoff, *Astronomy: From the Earth to the Universe* (Belmont, CA: Thomson Learning, 2002), pp. 397–409; and National Space Science Data Center (NSSDC), "Sun Fact Sheet," http://nssdc.gsfc.nasa.gov/planetary/factsheet/sunfact.html (accessed September 1, 2004).

*Because the density of water is 1.0 g/cm^3, it is often used as a standard when comparing the density of different objects. For example, the Sun is, on average, 1.4 times denser than water (William L. Masterton, Emil J. Slowinski, and Conrad L. Stanitski, *Chemical Principles*, 5th ed. [Philadelphia: Saunders College Publishing, 1981], p. 17).

THE PLANETS—MERCURY

Average Distance from the Sun .... 0.39 AU
35,950,000 miles (57,900,000 kilometers)

Equatorial Diameter .... 3,030 miles (4,880 kilometers)

Average Temperatures .... Day: 660°F (350°C)
Night: −270°F (−170°C)

Length of One Day .... 59 Earth days

Length of One Year .... 88 Earth days

Atmospheric Composition .... No atmosphere

Mass .... 7.26×10^{23} pounds (3.30×10^{23} kilograms)

Average Density .... 5.4 g/cm^3

Number of Moons .... No Moons

Information gathered from the following sources: Roger A. Freedman and William J. Kaufmann III, *Universe*, 6th ed. (New York: W. H. Freeman and Company, 2002), pp. 225–37; Jay M. Pasachoff, *Astronomy: From the Earth to the Universe* (Belmont, CA: Thomson Learning, 2002), pp. 175–87; and NSSDC, "Mercury Fact Sheet," http://nssdc.gsfc.nasa.gov/planetary/factsheet/mercuryfact.html (accessed September 1, 2004).

THE PLANETS—VENUS

Average Distance from the Sun .... 0.72 AU

67,190,000 miles (108,200,000 kilometers)

Equatorial Diameter .... 7,515 miles (12,102 kilometers)

Average Temperature .... 900°F (480°C)

Length of One Day .... 243 Earth days

Length of One Year .... 224.7 Earth days

Atmospheric Composition .... 97% Carbon Dioxide

3% Nitrogen

Mass .... 1.07×10^{25} pounds (4.87×10^{24} kilograms)

Average Density .... 5.2 g/cm^3

Number of Moons .... No Moons

Information gathered from the following sources: Roger A. Freedman and William J. Kaufmann III, *Universe*, 6th ed. (New York: W. H. Freeman and Company, 2002), pp. 240–56; Jay M. Pasachoff, *Astronomy: From the Earth to the Universe* (Belmont, CA: Thomson Learning, 2002), pp. 191–205; and NSSDC, "Venus Fact Sheet," http://nssdc.gsfc.nasa.gov/planetary/factsheet/venusfact.html (accessed September 1, 2004).

THE PLANETS—EARTH

Average Distance from the Sun.............. 1.0 AU

92,900,000 miles (149,600,000 kilometers)

Equatorial Diameter .... 7,921 miles (12,756 kilometers)

Average Temperature .... 70°F (20°C)

Length of One Day .... 23 hours, 56 minutes

Length of One Year............................... 365.25 Earth days

Atmospheric Composition .... 77% Nitrogen

21% Oxygen

2% other gases

Mass .... 1.31×10^{25} pounds (5.97×10^{24} kilograms)

Average Density .... 5.5 g/cm^3

Number of Moons 1

Information gathered from the following sources: Roger A. Freedman and William J. Kaufmann III, *Universe*, 6th ed. (New York: W. H. Freeman and Company, 2002), pp. 179–201; Jay M. Pasachoff, *Astronomy: From the Earth to the Universe* (Belmont, CA: Thomson Learning, 2002), pp. 135–51; and NSSDC, "Earth Fact Sheet," http://nssdc.gsfc.nasa.gov/planetary/factsheet/earthfact.html (accessed September 1, 2004).

THE PLANETS—MARS

Average Distance from the Sun.............. 1.52 AU

141,500,000 miles (227,900,000 kilometers)

Equatorial Diameter .... 4,219 miles (6,794 kilometers)

Average Temperatures........................... Hottest Day: 70°F (20°C)

Coldest Night: −220°F (−140°C)

Length of One Day .... 24 hours, 39 minutes

Length of One Year................................ 687 Earth days

Atmospheric Composition .... 95% Carbon Dioxide

3% Nitrogen

2% other gases

Mass .... 1.41×10^{24} pounds (6.42×10^{23} kilograms)

Average Density .... 3.9 g/cm^3

Number of Moons 2

Information gathered from the following sources: Roger A. Freedman and William J. Kaufmann III, *Universe*, 6th ed. (New York: W. H. Freeman and Company, 2002), pp. 260–78; Jay M. Pasachoff, *Astronomy: From the Earth to the Universe* (Belmont, CA: Thomson Learning, 2002), pp. 207–26; and NSSDC, "Mars Fact Sheet," http://nssdc.gsfc.nasa.gov/planetary/factsheet/marsfact.html (accessed September 1, 2004).

THE PLANETS—JUPITER

Average Distance from the Sun .... 5.2 AU

483,300,000 miles (778,300,000 kilometers)

Equatorial Diameter .... 88,793 miles (142,984 kilometers)

Average Temperature .... −166°F (−110°C)

Length of One Day .... Equator: 9 hours, 50 minutes

Poles: 9 hours, 55 minutes

Length of One Year .... 11.86 Earth years

Atmospheric Composition .... 82% Hydrogen

17% Helium

1% other gases

Mass .... 4.18×10^{27} pounds (1.899×10^{27} kilograms)

Average Density .... 1.3 g/cm^3

Number of Moons 63

Information gathered from the following sources: Roger A. Freedman and William J. Kaufmann III, *Universe*, 6th ed. (New York: W. H. Freeman and Company, 2002), pp. 283–97; Jay M. Pasachoff, *Astronomy: From the Earth to the Universe* (Belmont, CA: Thomson Learning, 2002), pp. 229–49; and NSSDC, "Jupiter Fact Page," http://nssdc.gsfc.nasa.gov/planetary/factsheet/jupiterfact.html (accessed September 1, 2004).

THE PLANETS—SATURN

Average Distance from the Sun .... 9.55 AU
887,400,000 miles (1,429,000,000 kilometers)

Equatorial Diameter .... 74,853 miles (120,536 kilometers)

Average Temperature .... −292°F (−180°C)

Length of One Day .... 10 hours, 14 minutes

Length of One Year .... 29.5 Earth years

Atmospheric Composition .... 88% Hydrogen
11% Helium
1% other gases

Mass .... 1.25×10^{27} pounds (5.69×10^{26} kilograms)

Average Density .... 0.69 g/cm^3

Number of Moons 33

Information gathered from the following sources: Roger A. Freedman and William J. Kaufmann III, *Universe*, 6th ed. (New York: W. H. Freeman and Company, 2002), pp. 322–37; Jay M. Pasachoff, *Astronomy: From the Earth to the Universe* (Belmont, CA: Thomson Learning, 2002), pp. 253–68; and NSSDC, "Saturn Fact Page," http://nssdc.gsfc.nasa.gov/planetary/factsheet/saturnfact.html (accessed September 1, 2004).

THE PLANETS—URANUS

Average Distance from the Sun .... 19.22 AU

1,785,000,000 miles (2,875,000,000 kilometers)

Equatorial Diameter .... 31,744 miles (51,118 kilometers)

Average Temperature .... −364°F (−220°C)

Length of One Day .... 17 hours, 14 minutes

Length of One Year............................... 84 Earth years

Atmospheric Composition .... 84% Hydrogen

14% Helium

2% Methane

Mass .... 1.91×10^{26} lb

8.66×10^{25} kg

Average Density .... 1.3 g/cm^3

Number of Moons 27

Information gathered from the following sources: Roger A. Freedman and William J. Kaufmann III, *Universe*, 6th ed. (New York: W. H. Freeman and Company, 2002), pp. 342–56; Jay M. Pasachoff, *Astronomy: From the Earth to the Universe* (Belmont, CA: Thomson Learning, 2002), pp. 271–83; and NSSDC, "Uranus Fact Sheet," http://nssdc.gsfc.nasa.gov/planetary/factsheet/uranusfact.html (accessed September 1, 2004).

THE PLANETS—NEPTUNE

Average Distance from the Sun .... 30.1 AU

2,982,000,000 miles (4,504,000,000 kilometers)

Equatorial Diameter .... 30,757 miles (49,528 kilometers)

Average Temperature .... −360°F (−218°C)

Length of One Day .... 16 hours, 17 minutes

Length of One Year .... 165 Earth years

Atmospheric Composition .... 84% Hydrogen

14% Helium

2% other gases

Mass .... 2.26×10^{26} pounds (1.03×10^{26} kilograms)

Average Density .... 1.6 g/cm^3

Number of Moons 13

Information gathered from the following sources: Roger A. Freedman and William J. Kaufmann III, *Universe*, 6th ed. (New York: W. H. Freeman and Company, 2002), pp. 342–56; Jay M. Pasachoff, *Astronomy: From the Earth to the Universe* (Belmont, CA: Thomson Learning, 2002), pp. 287–300; and NSSDC, "Neptune Fact Sheet," http://nssdc.gsfc.nasa.gov/planetary/factsheet/neptunefact.html (accessed September 1, 2004).

THE PLANETS—PLUTO

Average Distance from the Sun.............. 39.5 AU

3,674,000,000 miles (5,916,000,000 kilometers)

Equatorial Diameter .... 1,430 miles (2,300 kilometers)

Average Temperature .... −387°F (−233°C)

Length of One Day .... 6 Earth days, 9 hours

Length of One Year.............................. 247.7 Earth years

Atmospheric Composition .... Traces of Nitrogen and Carbon Monoxide gases

Mass .... 2.86×10^{22} pounds (1.3×10^{22} kilograms)

Average Density .... 2.0 g/cm^3

Number of Moons 1

Information gathered from the following sources: Roger A. Freedman and William J. Kaufmann III, *Universe*, 6th ed. (New York: W. H. Freeman and Company, 2002), pp. 356–59; Jay M. Pasachoff, *Astronomy: From the Earth to the Universe* (Belmont, CA: Thomson Learning, 2002), pp. 303–12; and NSSDC, "Pluto Fact Sheet," http://nssdc.gsfc.nasa.gov/planetary/factsheet/plutofact.html (accessed September 1, 2004).

THE MOON

Distance from Earth........................... Average: 238,700 miles (384,400 kilometers)

Maximum: 251,800 miles (405,500 kilometers)

Minimum: 225,600 miles (363,300 kilometers)

Diameter .... 2,158 miles (3,476 kilometers)

Temperature .... Day: 266°F (130°C)

Night: −292°F (−180°C)

Rotation Rate .... New Moon to New Moon: 29.5 days

Relative to fixed stars: 27.3 days

Atmosphere .... None

Average Density .... 3.3 g/cm^3

Mass .... 1.62×10^{23} pounds (7.35×10^{23} kilograms)

Information gathered from the following sources: Roger A. Freedman and William J. Kaufmann III, *Universe*, 6th ed. (New York: W. H. Freeman and Company, 2002), pp. 206–21; Jay M. Pasachoff, *Astronomy: From the Earth to the Universe* (Belmont, CA: Thomson Learning, 2002), pp. 155–71; and NSSDC, "Moon Fact Sheet," http://nssdc.gsfc.nasa.gov/planetary/factsheet/moonfact.html (accessed September 22, 2004).

THE PHASES OF THE MOON

New Moon

 A New Moon sets with the Sun and cannot be seen from Earth because all of its sunlit side is facing away from our planet. With no moon in the sky, this is the best time to observe stars, star clusters, nebulae, and other faint objects.

Waxing Crescent

 A Waxing Crescent Moon is visible in the west southwestern sky just after the Sun sets. The Moon will be visible during the early evening, generally sinking below the horizon before midnight. The time between a Waxing Crescent and a First Quarter Moon is the best time to observe the Moon with a telescope.

First Quarter

 A First Quarter Moon rises above the eastern horizon when the Sun is directly overhead. When the Sun sets in the west, the Moon will be high overhead. A First Quarter Moon is a great sight through a telescope, especially along the terminator (the dividing line between dark and light). A First Quarter Moon generally sinks below the horizon around midnight.

Waxing Gibbous

 A Waxing Gibbous Moon is visible most of the night, rising in the east just before the Sun sets in the west. The term "Gibbous" describes the Moon when you can see more than half but not quite all of the Moon's surface.

Full Moon

 A Full Moon rises in the east as the Sun sets in the west and is visible all night. As the Sun rises in the east the next morning, the Moon will be sinking below the western horizon. Because of the angle of the sunlight striking the lunar surface, a Full Moon appears not only very bright in the sky, but also washed out and flat through a telescope. It's best to enjoy a Full Moon with your eyes alone.

Waning Gibbous

 A Waning Gibbous Moon rises in the east well after the Sun sets in the west and will light up the late night and early morning skies. As the Sun rises in the east the next morning, a Waning Gibbous Moon can still be seen low in the western skies.

Third Quarter

 A Third Quarter Moon rises late in the evening, generally around midnight, but can be seen throughout the morning in the west southwestern sky. A Third Quarter Moon sinks below the horizon around noon.

Waning Crescent

 A Waning Crescent Moon is visible in the east during the early morning hours, just before sunrise.

Then, the cycle (which takes about 29.5 days) starts all over again with a New Moon.

MOON TOTALS FOR EACH PLANET

Planet	Number of Moons
Mercury	0
Venus	0
Earth	1
Mars	2
Jupiter	63
Saturn	33
Uranus	27
Neptune	13
Pluto	1

MOON NAMES

The moons are listed by planet, starting with the closest moon to each planet and moving outward.

Moons of Earth
    Luna

Moons of Mars
    Phobos, Deimos

Moons of Jupiter
    Metis, Adrastea, Amalthea, Thebe, Io, Europa, Ganymede, Callisto, Themisto, Leda, Himalia, Lysithea, Elara, Euporie, Euanthe, Harpalyke, Praxidike, Orthosie, Hermippe, Iocaste, Ananke, Thyone, Pasithee, Kale, Chaldene, Eurydome, Isonoe, Erinome, Taygete, Carme, Aitne, Kalyke, Pasiphae, Sponde, Megaclite, Sinope, Callirrho, Autonoe

National Space Science Data Center (NSSDC), "Jovian Satellite Fact Sheet," http://nssdc.gsfc.nasa.gov/planetary/factsheet/joviansatfact.html (accessed September 1, 2004); NSSDC, "Saturnian Satellite Fact Sheet," http://nssdc.gsfc.nasa.gov/planetary/factsheet/saturniansatfact.html (accessed September 1, 2004); NSSDC, "Uranian Satellite Fact Sheet," http://nssdc.gsfc.nasa.gov/planetary/factsheet/uraniansatfact.html (accessed September 1, 2004); NSSDC, "Neptunian Satellite Fact Sheet," http://nssdc.gsfc.nasa.gov/planetary/factsheet/neptuniansatfact.html (accessed September 1, 2004).

Moons of Saturn

Pan, Atlas, Prometheus, Pandora, Epimetheus, Janus, Mimas, Enceladus, Tethys, Calypso, Telesto, Dione, Helene, Rhea, Titan, Hyperion, Iapetus, Kiviuq, Ijiraq, Phoebe, Paaliaq, Skadi, Albiorix, Erriapo, Siarnaq, Tarvos, Mundilfari, Suttung, Thrym, Ymir

Moons of Uranus

Cordelia, Ophelia, Bianca, Cressida, Desdemona, Juliet, Portia, Rosalind, Belinda, Puck, Miranda, Ariel, Umbriel, Titania, Oberon, Caliban, Stephano, Trinculo, Sycorax, Prospero, Setebos

Moons of Neptune

Naiad, Thalassa, Despina, Galatea, Larissa, Proteus, Triton, Nereid

Moons of Pluto

Charon

THE BIGGEST MOONS
OF THE SOLAR SYSTEM

Rank	Moon	Diameter (in miles)	Diameter (in kilometers)	Planet
1	Ganymede	3,268	5,262	Jupiter
2	Titan	3,198	5,150	Saturn
3	Callisto	2,981	4,800	Jupiter
4	Io	2,254	3,630	Jupiter
5	Luna	2,158	3,476	Earth
6	Europa	1,949	3,138	Jupiter
7	Triton	1,677	2,700	Neptune
8	Titania	981	1,580	Uranus
9	Rhea	950	1,530	Saturn
10	Oberon	944	1,520	Uranus
11	Iapetus	907	1,460	Saturn
12	Charon	739	1,190	Pluto
13	Umbriel	727	1,170	Uranus
14	Ariel	720	1,160	Uranus
15	Dione	696	1,120	Saturn
16	Tethys	658	1,060	Saturn
17	Enceladus	310	500	Saturn
18	Miranda	292	470	Uranus
19	Proteus	258	415	Neptune
20	Mimas	243	392	Saturn
21	Nereid	211	340	Neptune
22	Phoebe	137	220	Saturn
23	Larissa	118	190	Neptune
24	Galatea	112	180	Neptune
25	Puck	93	150	Uranus
25	Despina	93	150	Neptune

(For more information, see the following questions: "Does every planet have a moon?" page 53, "Which planet has the most moons?" page 54, and "Which planet has the largest moon?" page 54.)

Information compiled from the sources listed on previous Sun, Moon, and planets fact pages.

LUNAR ECLIPSES FOR 2005–2020

Date	Time of Maximum Coverage*	Eclipse Type	Eclipse Duration†	Where Visible
2005 Oct 17	12:03 UT (8:03 AM EDT)	Partial	00h58m	Asia, Australia, Pacific, North America
2006 Sep 07	18:51 UT	Partial	01h33m	Europe, Africa, Asia, Australia
2007 Mar 03	23:21 UT (6:21 PM EST)	Total	03h42m, **01h14m**	Americas, Europe, Africa, Asia
2007 Aug 28	10:37 UT (6:37 AM EDT)	Total	03h33m, **01h31m**	eastern Asia, Australia, Pacific, Americas
2008 Feb 21 (2008 Feb 20)	03:26 UT (10:26 PM EST)	Total	03h26m, **00h51m**	central Pacific, Americas, Europe, Africa
2008 Aug 16	21:10 UT	Partial	03h09m	South America, Europe, Africa, Asia, Australia
2009 Dec 31	19:23 UT	Partial	01h02m	Europe, Africa, Asia, Australia
2010 Jun 26	11:38 UT (7:38 AM EDT)	Partial	02h44m	eastern Asia, Australia, Pacific, western Americas
2010 Dec 21	08:17 UT (3:17 AM EST)	Total	03h29m, **01h13m**	eastern Asia, Australia, Pacific, Americas, Europe

Fred Espenak, "Lunar Eclipse Page," NASA Goddard Space Flight Center, http://sunearth.gsfc.nasa.gov/eclipse/lunar.html (accessed September 9, 2004).

2011 Jun 15	20:12 UT	Total	03h40m, **01h41m**	South America, Europe, Africa, Asia, Australia
2011 Dec 10	14:32 UT (9:32 AM EST)	Total	03h33m, **00h52m**	Europe, eastern Africa, Asia, Australia, Pacific, North America
2012 Jun 04	11:03 UT (7:03 AM EDT)	Partial	02h08m	Asia, Australia, Pacific, Americas
2013 Apr 25	20:07 UT	Partial	00h32m	Europe, Africa, Asia, Australia
2014 Apr 15	07:46 UT (3:46 AM EDT)	Total	03h35m, **01h19m**	Australia, Pacific, Americas
2014 Oct 08	10:54 UT (6:54 AM EDT)	Total	03h20m, **01h00m**	Asia, Australia, Pacific, Americas
2015 Apr 04	12:00 UT (7:00 AM EST)	Total	03h30m, **00h12m**	Asia, Australia, Pacific, Americas
2015 Sep 28 (2015 Sep 27)	02:47 UT (10:47 PM EDT)	Total	03h21m, **01h13m**	eastern Pacific, Americas, Europe, Africa, western Asia
2017 Aug 07	18:20 UT	Partial	01h57m	Europe, Africa, Asia, Australia
2018 Jan 31	13:30 UT (8:30 AM EST)	Total	03h23m, **01h17m**	Asia, Australia, Pacific, western North America
2018 Jul 27	20:22 UT	Total	03h55m, **01h44m**	South America, Europe, Africa, Asia, Australia
2019 Jan 21	05:12 UT (12:12 AM EST)	Total	03h17m, **01h03m**	central Pacific, Americas, Europe, Africa
2019 Jul 16	21:31 UT	Partial	02h59m	South America, Europe, Africa, Asia, Australia
2020	No partial or total lunar eclipses occur in 2020			

*Times given in Universal Time (UT) and Eastern Standard Time (EST) or Eastern Daylight Time (EDT) when appropriate.

†First time indicates duration of eclipse from beginning to end. For total eclipses, second time (in bold letters) indicates time of total coverage.

SOLAR ECLIPSES FOR 2005–2020

Date	Eclipse Type	Time of Maximum Coverage*	Duration of Total Phase	Partial eclipse visible from:	Total or Annular eclipse visible from:
2005 Apr 08	Annular/ Total	20:36 UT (4:36 PM EDT)	00m42s	New Zealand, Americas	southern Pacific, Panama, Colombia, Venezuela
2005 Oct 03	Annular	10:32 UT	04m32s	Europe, Africa, southern Asia	Portugal, Spain, Libya, Sudan, Kenya
2006 Mar 29	Total	10:11 UT	04m07s	Africa, Europe, western Asia	central Africa, Turkey, Russia
2006 Sep 22	Annular	11:40 UT	07m09s	South America, western Africa, Antarctica	Guyana, Suriname, French Guiana, southern Atlantic
2007 Mar 19	Partial	02:32 UT	—	Asia, Alaska	—
2007 Sep 11	Partial	12:31 UT	—	South America, Antarctica	—
2008 Feb 07	Annular	03:55 UT	02m12s	Antarctica, eastern Australia, New Zealand	Antarctica
2008 Aug 01	Total	10:21 UT (6:21 AM EDT)	02m27s	northeastern North America, Europe, Asia	northern Canada, Greenland, Siberia, Mongolia, China

Fred Espenak, "Solar Eclipse Page," NASA Goddard Space Flight Center, http://sunearth.gsfc.nasa.gov/eclipse/solar.html (accessed July 29, 2004).

2009 Jan 26	Annular	07:58 UT	07m54s	southern Africa, Antarctica, southeast Asia, Australia	southern Indian Ocean, Sumatra, Borneo
2009 Jul 22	Total	02:35 UT	06m39s	eastern Asia, Pacific Ocean, Hawaii	India, Nepal, China, central Pacific
2010 Jan 15	Annular	07:06 UT	11m08s	Africa, Asia	central Africa, India, Myanmar, China
2010 Jul 11	Total	19:33 UT	05m20s	southern South America	southern Pacific, Easter Island, Chile, Argentina
2011 Jan 04	Partial	08:50 UT	—	Europe, Africa, central Asia	—
2011 Jun 01	Partial	21:16 (5:16 PM EDT)	—	eastern Asia, northern North America, Iceland	—
2011 Jul 01	Partial	08:38 UT	—	southern Indian Ocean	—
2011 Nov 25	Partial	06:20 UT	—	southern Africa, Antarctica, Tasmania, New Zealand	—
2012 May 20	Annular	23:53 UT (7:52 PM EDT)	05m46s	Asia, Pacific, North America	China, Japan, Pacific, western United States
2012 Nov 13	Total	22:12 UT	04m02s	Australia, New Zealand, southern Pacific, southern South America	northern Australia, southern Pacific
2013 May 10	Annular	00:25 UT	06m03s	Australia, New Zealand, central Pacific	northern Australia, Solomon Islands, central Pacific
2013 Nov 03	Annular/ Total	12:46 UT (7:46 AM EST)	01m40s	eastern Americas, southern Europe, Africa	Atlantic, central Africa
2014 Apr 29	Annular	06:03 UT	Non-Central	southern Indian Ocean, Australia, Antarctica	Antarctica
2014 Oct 23	Partial	21:44 UT (5:44 PM EDT)	—	northern Pacific, North America	—

2015 Mar 20	Total	09:45 UT	02m47s	Iceland, Europe, northern Africa, northern Asia	northern Atlantic, Faeroe Islands, Svalbard
2015 Sep 13	Partial	06:54 UT	—	southern Africa, southern Indian Ocean, Antarctica	—
2016 Mar 09	Total	01:57 UT	04m09s	eastern Asia, Australia, Pacific	Sumatra, Borneo, Sulawesi, Pacific
2016 Sep 01	Annular	09:07 UT	03m06s	Africa, Indian Ocean	Atlantic, central Africa, Madagascar, Indian Ocean
2017 Feb 26	Annular	14:53 UT	00m44s	southern South America, Atlantic, Africa, Antarctica	Pacific, Chile, Argentina, Atlantic, Africa
2017 Aug 21	Total	18:25 UT (2:25 PM EDT)	02m40s	North America, northern South America	northern Pacific, United States, southern Atlantic
2018 Feb 15	Partial	20:51 UT	—	Antarctica, southern South America	—
2018 Jul 13	Partial	03:01 UT	—	southern Australia	—
2018 Aug 11	Partial	09:46 UT	—	northern Europe, northeast Asia	—
2019 Jan 06	Partial	01:41 UT	—	northeast Asia, northern Pacific	—
2019 Jul 02	Total	19:23 UT	04m33s	southern Pacific, South America	southern Pacific, Chile, Argentina
2019 Dec 26	Annular	05:17 UT	03m40s	Asia, Australia	Saudi Arabia, India, Sumatra, Borneo
2020 Jun 21	Annular	06:40 UT	00m38s	Africa, southeast Europe, Asia	central Africa, southern Asia, China, Pacific
2020 Dec 14	Total	16:13 UT	02m10s	Pacific, southern South America, Antarctica	southern Pacific, Chile, Argentina, southern Atlantic

*Times given in Universal Time (UT) and Eastern Standard Time (EST) or Eastern Daylight Time (EDT) when appropriate.

METEOR SHOWERS

The following is a list of the best meteor showers that can be seen throughout the year.

Shower Name	When Visible	Peak Date (best night for viewing)	Hourly Rate
Quadrantid	January 1–5	January 3	40
Lyrid	April 20–24	April 22	15
η (Eta) Aquarid	May 3–6	May 5	20
δ (Delta) Aquarid	July 29–31	July 30	35
Perseid	August 10–14	August 12	50
Orionid	October 20–22	October 21	30
Taurid	November 3–6	November 5	15
Leonid	November 13–19	November 17	15
Geminid	December 10–15	December 13	50
Ursid	December 21–23	December 22	15

Information derived from the following sources: Roger A. Freedman and William J. Kaufmann III, *Universe*, 6th ed. (New York: W. H. Freeman and Company, 2002), p. 383; Jay M. Pasachoff, *Astronomy: From the Earth to the Universe* (Belmont, CA: Thomson Learning, 2002), p. 357; Jeffrey Bennett et al., *The Solar System: The Cosmic Perspective*, 3rd ed. (San Francisco: Addison Wesley, 2004), p. 386; Harvard-Smithsonian Center for Astrophysics, "Major Annual Meteor Showers," http://cfa-www.harvard.edu/cfa/ep/meteor/shower1.html (accessed March 27, 2001); and National Weather Service Forecast Office, "Meteors and Meteor Showers," National Oceanic and Atmospheric Administration, http://www.crh.noaa.gov/fsd/astro/meteor.htm (accessed June 22, 2004).

Tips for viewing a meteor shower:

- Get away from city lights. (The darker the sky, the better your chance of seeing faint meteors.)

- Choose an observing location away from trees or buildings. (The more sky you can see, the better your opportunities of seeing a meteor.)

- Bring along comfort items like blankets, lawn chairs, insect repellant, and snacks. (Because a meteor shower will last all night, it is best to be prepared!)

- Don't worry about telescopes or binoculars. (A meteor shower is best observed with your eyes alone.)

- The best time to observe meteor showers (with regards to Moon phases) is when the Moon is between Third Quarter and New Moon. A bright Moon in the sky can drown out all but the brightest meteors. Moon phases can be found in the weather section of your local paper.

- The best time to observe a meteor shower is between midnight and dawn, looking toward the eastern horizon. (However, if you can't stay up that late, it's still worth looking, whenever you get a chance.)

THE TWENTY-FIVE BRIGHTEST STARS

Rank	Star Name	Constellation	Magnitude* (Brightness)
1	Sirius	Canis Major	−1.44
2	Canopus	Carina	−0.62
3	Alpha Centauri	Centaurus	−0.28
4	Arcturus	Boötes	−0.05
5	Vega	Lyra	0.03
6	Capella	Auriga	0.08
7	Rigel	Orion	0.18
8	Procyon	Canis Minor	0.40
9	Achernar	Eridanus	0.45
10	Betelgeuse	Orion	0.45
11	Hadar	Centaurus	0.61
12	Altair	Aquila	0.76
13	Acrux	Crux	0.77
14	Aldebaran	Taurus	0.87
15	Spica	Virgo	0.98
16	Antares	Scorpius	1.06
17	Pollux	Gemini	1.16
18	Fomalhaut	Piscis Australis	1.17
19	Deneb	Cygnus	1.25
20	Mimosa	Crux	1.25
21	Regulus	Leo	1.36
22	Adhara	Canis Major	1.50
23	Castor	Gemini	1.58
24	Gacrux	Crux	1.59
25	Shaula	Scorpius	1.62

*Astronomers use a magnitude scale to rank a star's apparent brightness in the sky. The brighter the star, the smaller the magnitude. The brightest star in the night sky, Sirius, is rated at magnitude −1.44. The faintest stars that can be seen with the naked eye are around magnitude 5.5.

Information derived from the following sources: Millennium Star Atlas, "50 Brightest Stars," Sky Publishing Corp., Hipparcos, http://astro.estec.esa.nl/Hipparcos/msa-tab4.html (accessed September 10, 1999); "The Brightest Stars," Students for the Exploration and Development of Space, http://www.seds.org/Maps/Stars_en/ (accessed January 11, 1998); Cosmobrain Astronomy, "The 50 Brightest Stars," http://www.cosmobrain.com/ cosmobrain/res/brightstar.html (accessed September 22, 2004); and Gaston Lamontagne, "50 Brightest Stars," Royal Astronomical Society of Canada, Ottawa, http://ottawa.rasc.ca/astronomy/brightest_stars/brightest _stars.html (accessed March 23, 2003).

THE TWENTY-FIVE CLOSEST STARS

Rank	Star Name	Constellation	Distance (in light-years)
1	Our Sun	—	—
2	Alpha Centauri (Proxima Centauri)	Centaurus	4.2
3	Alpha Centauri (Rigel Kent A)	Centaurus	4.4
4	Alpha Centauri (Rigel Kent B)	Centaurus	4.4
5	Barnard's Star	Ophiuchus	6.0
6	Wolf 359	Leo	7.7
7	Lalande 21185	Ursa Major	8.2
8	Sirius A	Canis Major	8.6
9	Sirius B	Canis Major	8.6
10	Luyten 726–8A	Cetus	8.9
11	Luyten 726–8B	Cetus	8.9
12	Ross 154	Sagittarius	9.4
13	Ross 248	Andromeda	10.4
14	Epsilon Eridani	Eridanus	10.7
15	Ross 128	Virgo	10.9
16	Luyten 789–6	Aquarius	11.2
17	Epsilon Indi	Indus	11.3
18	61 Cyg A	Cygnus	11.4
19	61 Cyg B	Cygnus	11.4
20	Procyon A	Canis Minor	11.4
21	Procyon B	Canis Minor	11.4
22	BD +59 deg 1915 A	Draco	11.5
23	BD +59 deg 1915 B	Draco	11.5
24	Groombridge 34 A	Andromeda	11.6
25	Groombridge 34 B	Andromeda	11.6

This chart is constantly changing as astronomers discover fainter stars using more sensitive equipment. Also note that because of the great distances involved, it is difficult to determine the precise distances to the stars.

Information derived from the following sources: Hipparcos, "The 150 Stars in the Hipparcos Catalog Closest to the Sun," http://astro.estec.esa.nl/Hipparcos/table361.html (accessed September 15, 2003); Cosmobrain Astronomy, "Nearest Stars," http://www.cosmobrain.com/cosmobrain/res/nearstar.html (accessed September 22, 2004); Chris Dolan, "The Nearest Stars, as Seen from the Earth," Washburn Observatory, University of Wisconsin, Madison, http://www.astro.wisc.edu/~dolan/constellations/extra/nearest.html (accessed March 23, 2003); and National Maritime Observatory, "30 Closest Stars," Royal Obseratory, Greenwich, http://www.nmm.ac.uk/site/request/setTemplate:singlecontent/contentTypeA/conWebDoc/contentId/410/navId/00500300f00g (accessed September 22, 2004).

THE STARS OF THE BIG DIPPER

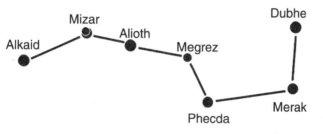

Common Name	Pronunciation	Stellar Designation	Distance in light-years	Magnitude* (Brightness)
Dubhe	DUBB-be	α—alpha (Ursa Majoris)	124 LY	1.81
Merak	ME-rack	β—beta (Ursa Majoris)	79 LY	2.34
Phecda	FECK-dah	γ—gamma (Ursa Majoris)	84 LY	2.41
Megrez	ME-grez	δ—delta (Ursa Majoris)	81 LY	3.32
Alioth	ALLEY-oth	ε—epsilon (Ursa Majoris)	81 LY	1.8
Mizar†	MY-zar	ζ—zeta (Ursa Majoris)	78 LY	2.23
Alkaid	al-KADE	η—eta (Ursa Majoris)	101 LY	1.85

*Astronomers use a magnitude scale to rank a star's apparent brightness in the sky. The brighter the star, the smaller the magnitude. The brightest star in the night sky, Sirius, is rated at magnitude −1.44. The faintest stars that can be seen with the naked eye are around magnitude 5.5.

†There is a second star, Alcor, that lies very close to Mizar. Some cultures used the pair to test eyesight—if you could see Alcor, you didn't need glasses! Other cultures referred to the pair as the horse and rider. The two stars are not connected to each other in any way. It's just that, from our vantage point on Earth, the two stars appear close together. See if you can make out Alcor the next time you look at the Big Dipper.

(For more information, see "What is a constellation?" on page 134, and "How can I find a constellation?" on page 136.)

Information derived from the following sources: Robert Burnham Jr., *Burnham's Celestial Handbook: An Observer's Guide to the Universe beyond the Solar System*, 3 vols. (New York: Dover Publishing, 1978), 3:1949–57; Software Bisque, *TheSky* Astronomy Software (Golden, CO: Software Bisque, 1998); and H. J. P. Arnold, P. Doherty, and P. Moore, *Photographic Atlas of the Stars* (Waukesha, WI: Kalmbach Publishing, 1997), 113.l

THE CONSTELLATIONS
(IN ALPHABETICAL ORDER)

Constellation	Description	Visible in Evening Sky	Hemisphere
Andromeda	The Maiden	October–December	North
Antlia	The Air Pump	April–June	South
Apus	The Bee	May–August	South
Aquarius	The Water Carrier	September–December	Both
Aquila	The Eagle	August–October	North
Ara	The Alter	July–September	South
Aries	The Ram	October–February	Both
Auriga	The Charioteer	January–March	North
Boötes	The Herdsman	June–August	North
Caelum	The Sculptor's Tool	January–February	South
Camelpardalis	The Giraffe	December–February	North
Cancer	The Crab	March–May	Both
Canes Venatici	The Hunting Dogs	May–July	North
Canis Major	The Big Dog	February–April	Both
Canis Minor	The Little Dog	February–April	Both
Capricornus	The Sea Goat	September–November	Both
Carina	The Keel	March–May	South
Cassiopeia	The Queen	September–December	North
Centaurus	The Centaur	May–July	South
Cepheus	The King	September–December	North
Cetus	The Sea Monster	October–December	Both

Information derived from the following sources: Robert Burnham Jr., *Burnham's Celestial Handbook: An Observer's Guide to the Universe beyond the Solar System*, 3 vols. (New York: Dover Publishing, 1978), 3: 2131–33; David Levy, *Skywatching: A Nature Company Guide* (New York: Time Life Books, 2000); Mark R. Chartrand, *The Audubon Society Field Guide to the Night Sky* (New York: Alfred A. Knopf, 1991); Roy A. Gallant, *The Constellations: How They Came to Be* (New York: Four Winds Press, 1991); and Edmund Scientific Company, *Edmund MAG 5 Star Atlas* (Barrington, NJ: Edmund Scientific, 1980), pp. 19–29.

Constellation	Description	Visible in Evening Sky	Hemisphere
Chamaeleon	The Chameleon	year round	South
Circinus	The Compass	July–September	South
Columba	The Dove	January–March	South
Coma Berenices	Berenices' Hair	May–July	North
Corona Australis	The Southern Crown	August–October	South
Corona Borealis	The Northern Crown	July–September	North
Corvus	The Crow	April–June	Both
Crater	The Cup	April–June	Both
Crux	The Southern Cross	April–June	South
Cygnus	The Swan	August–November	North
Delphinus	The Dolphin	August–October	North
Dorado	The Swordfish	December–February	South
Draco	The Dragon	July–September	North
Equuleus	The Foal	August–October	Both
Eridanus	The River	December–February	Both
Fornax	The Furnace	December–February	South
Gemini	The Twins	January–April	Both
Grus	The Crane	September–November	South
Hercules	Hercules	July–September	North
Horologium	The Clock	December–February	South
Hydra	The Water Snake	March–August	Both
Hydrus	The Little Snake	year round	South
Indus	The Indian	September–November	South
Lacerta	The Lizard	September–November	North
Leo	The Lion	March–June	Both
Leo Minor	The Little Lion	March–May	Both
Lepus	The Rabbit	January–March	Both
Libra	The Scales	June–July	Both
Lupus	The Wolf	June–July	Both
Lynx	The Lynx	February–April	North
Lyra	The Harp	June–September	North
Mensa	The Table	year round	South
Microscopium	The Microscope	September–October	South
Monoceros	The Unicorn	February–March	Both
Musca	The Fly	year round	South
Norma	The Level and Square	July–August	South
Octans	The Octant	year round	South
Ophiuchus	The Serpent Bearer	July–August	Both
Orion	The Hunter	January–March	Both
Pavo	The Peacock	September–October	South
Pegasus	The Flying Horse	September–November	Both
Perseus	Perseus	October–December	North
Phoenix	The Phoenix	October–November	South
Pictor	The Painter	January–March	South
Pisces	The Fish	October–December	Both
Piscis Australis	The Southern Fish	October–November	South
Puppis	The Stern	March–April	South
Pyxis	The Mariner's Compass	March–April	South

Constellation	Description	Visible in Evening Sky	Hemisphere
Reticulum	The Net	December–January	South
Sagitta	The Arrow	August–October	North
Sagittarius	The Archer	July–August	Both
Scorpius	The Scorpion	June–August	Both
Sculptor	The Sculptor	November–December	South
Scutum	The Shield	August–October	Both
Serpens	The Serpent	July–August	Both
Sextans	The Sextant	April–June	Both
Taurus	The Bull	October–December	Both
Telescopium	The Telescope	August–September	South
Triangulum	The Triangle	October–December	North
Triangulum Australe	The Southern Triangle	year round	South
Tucana	The Toucan	September–November	South
Ursa Major	The Big Bear	year round	North
Ursa Minor	The Little Bear	year round	North
Vela	The Sails	March–April	South
Virgo	The Virgin	April–June	Both
Volans	The Flying Fish	year round	South
Vulpecula	The Fox	August–October	North

ASTRONOMY AND SPACE-RELATED WEB SITES

Below are just a few of the many wonderful Web sites available to anyone interested in astronomy, outer space, and space exploration. Many of these sites were used during the research phase of this book. Please keep in mind that with the ever-changing nature of the Web, the status of a particular Web page might change overnight. Moreover, some Web sites (not given here) contain inaccurate information. This section is merely to get you pointed in the right direction. It is by no means a complete representation of all the reliable Web sites out there.

Space Exploration—Human and Robotic

National Aeronautics and Space Administration (NASA)
http://www.nasa.gov

NASA's Human Space Flight Page
http://spaceflight.nasa.gov/home/index.html

Jet Propulsion Laboratory
http://www.jpl.nasa.gov

Kennedy Space Center
http://www.ksc.nasa.gov

Johnson Space Center
http://www.jsc.nasa.gov

Space Shuttle

Space Shuttle Page (Kennedy Space Center)
http://www.ksc.nasa.gov/shuttle

Space Shuttle Page (NASA's Human Space Flight)
http://spaceflight.nasa.gov/shuttle

Space Shuttle Launch Archive
http://science.ksc.nasa.gov/shuttle/missions/missions.html

Space Shuttle Online Launch Schedule
http://www-pao.ksc.nasa.gov/kscpao/schedule/schedule.htm

International Space Station

International Space Station
http://www.shuttlepresskit.com/ISS_OVR/

International Space Station (Boeing)
http://www.boeing.com/defense-space/space/spacestation/sitemap.html

International Space Station (NASA's Human Space Flight)
http://spaceflight.nasa.gov/station

International Space Station (European Space Agency)
http://www.esa.int/export/esaHS/iss.html

Apollo Program

Project Apollo (Kennedy Space Center)
http://science.ksc.nasa.gov/history/apollo/apollo.html

The Apollo Program: 1963–1972
http://nssdc.gsfc.nasa.gov/planetary/lunar/apollo.html

Johnson Space Center History Portal
http://www.jsc.nasa.gov/history/

Robotic Missions (Just a Few)

Cassini Mission to Saturn
http://saturn.jpl.nasa.gov/home/index.cfm

Mars Exploration Rover Mission (*Spirit* and *Opportunity*)
http://marsrovers.jpl.nasa.gov/home

MESSENGER Mission to Mercury
http://messenger.jhuapl.edu

Stardust Mission to a Comet
http://stardust.jpl.nasa.gov

New Horizons Pluto–Kuiper Belt Mission
http://www.pluto.jhuapl.edu

For more robotic missions, visit

Jet Propulsion Laboratory Web site
http://www.jpl.nasa.gov

and

Johns Hopkins University Applied Physics Laboratory
http://www.jhuapl.edu

Space Images

Johnson Space Center Digital Image Collection
http://images.jsc.nasa.gov/

Planetary Photojournal (NASA)
http://photojournal.jpl.nasa.gov/index.html

Hubble Space Telescope Gallery
http://hubblesite.org/gallery/

Hubble Space Telescope—Wide Field and Planetary Camera 2
http://www.jpl.nasa.gov/images/wfpc/

Astronomy and Space Information

National Space Science Data Center
http://nssdc.gsfc.nasa.gov

International Astronomical Union
http://www.iau.org/

Space Observatories

Hubble Space Telescope
http://hubblesite.org/

Chandra X-Ray Observatory
http://chandra.harvard.edu/index.html

Spitzer Space Telescope
http://www.spitzer.caltech.edu

Astronomical Events

Eclipses
http://sunearth.gsfc.nasa.gov/eclipse/eclipse.html

Solar Eclipses
http://sunearth.gsfc.nasa.gov/eclipse/solar.html

Lunar Eclipses
http://sunearth.gsfc.nasa.gov/eclipse/lunar.html

The American Meteor Society
http://www.amsmeteors.org/

International Meteor Organization
http://www.imo.net/index.html

Comet Observation Home Page
http://encke.jpl.nasa.gov/

International Comet Quarterly
http://cfa-www.harvard.edu/icq/cometobs.html

Ways to Get Involved in Astronomy and Space

The Planetary Society
http://planetary.org

Astronomical Society of the Pacific
http://www.astrosociety.org

Astronomical League
http://www.astroleague.org

International Dark Sky Association
http://www.darksky.org

Association of Lunar and Planetary Observers
http://www.lpl.arizona.edu/alpo

Heavens-Above
http://www.heavens-above.com

PART 3

A BRIEF HISTORY OF LUNAR AND PLANETARY EXPLORATION

*S*tar Trek (and just about every other science fiction show or movie) makes space travel look easy—except of course for the occasional "warp core breach" or "transporter malfunction." The reality of space travel, however, is quite different.

Building a lightweight spacecraft tough enough to withstand being bounced around during launch that can also survive the rapid transition from Earth's warm, dense atmosphere to the frigid vacuum of outer space may seem difficult enough. That same spacecraft must also be able to survive days, months, or even years of travel (depending on its destination) in the harsh environment of interplanetary space before it reaches its target. If the spacecraft survives and reaches its destination with all of its instruments still working, it then must be able to collect images and scientific data and transmit that information back to Earth.

The above description is an extremely simplified version of what actually happens during a real-life space mission. Because space travel is so difficult, there have been many failures, some catastrophic and tragic. However, there have been brilliant successes as well. The information these missions have provided has greatly increased our knowledge of the solar system and helped us gain a better understanding of our own planet. In addition, the technologies developed to make these space missions possible have affected our everyday lives. Everything from the satellites used to provide your daily weather forecast and worldwide news coverage to the computer on your desk or the digital watch on your arm has direct ties to a space program that began almost half a century ago.

The following brief history of lunar and planetary exploration contains all of the missions (the successes and the failures) that have been sent to the Moon and planets by all the spacefaring peoples: the United States, the former Soviet Union, Japan, China, and Europe. For the most part, these missions have been robotic in nature (i.e., no humans onboard). Only the nine manned Apollo missions that flew to the Moon are included here. All other human spaceflight missions have remained in orbit around our own planet.

The missions included in this book are sorted by their intended destination (the Moon, Mercury, Venus, etc.) and are listed in chronological order. Each entry includes the mission's name, launch date, arrival at its target (if appropriate), the sponsoring country, and a brief summary of the results. At the beginning of each chapter, there is also a list of interesting milestones to note as you read through the mission descriptions. A list of all the sources used when researching this section can be found beginning on page 307.

So, forget warp speed! Let's see what's really been achieved.

MISSIONS TO THE MOON

Our journey to the Moon begins in 1958 with the launch of *Pioneer 0*. Its mission lasted only seventy-seven seconds before the rocket's first stage exploded, but it marked the beginning of the United States' race to the Moon.

With *Pioneer 0*, we begin a voyage back in time to the era of the cold war and the intense rivalry between the United States and the Soviet Union. During the race to be first to the Moon, space missions were launched at an incredible rate, almost one a month. In the beginning, many of these missions failed, either on the launch pad or in Earth orbit. While many of these failures were catastrophic in nature, engineers learned from their mistakes and applied their newly acquired knowledge to the next generation of spacecraft. The pace of lunar exploration slowed dramatically after the Apollo missions in the early 1970s. Since then, various presidential space exploration initiatives have been proposed, including a return of humans to the Moon. It remains to be seen if any of these initiatives will really take us back to the Moon.

As you read through the missions, here are some interesting milestones to remember:

Apollo astronaut exploring the Moon.
Image credit: NASA

- ☄ the first spacecraft to fly by the Moon (*Pioneer 4*)
- ☄ the first spacecraft to crash into the Moon (*Luna 2*)
- ☄ the first controlled landing on the Moon (*Luna 9*)
- ☄ the mission that carried turtles and worms around the Moon and back to Earth (*Zond 5*)
- ☄ the first spacecraft to go into orbit around the Moon (*Luna 10*)
- ☄ the only intentional liftoff and landing back on the Moon (*Surveyor 6*)
- ☄ the first successful robotic sample return mission from the Moon (*Luna 16*)
- ☄ the first manned mission to land on the Moon (*Apollo 11*)
- ☄ the Apollo mission that was struck by lightning just minutes after liftoff (*Apollo 12*)
- ☄ the first remote-controlled rover on the Moon (*Lunokhod 1*)
- ☄ the first golf swing on the Moon (*Apollo 14*)
- ☄ the first manned mission to include a rover (*Apollo 15*)
- ☄ the mission that brought back the oldest samples of the Moon's surface (*Apollo 16*)
- ☄ the last manned mission to the Moon (*Apollo 17*)

THE MISSIONS

Pioneer 0

Launch Date: August 17, 1958

Sponsoring Country: USA

Type of Mission: Orbiter (designed to go into orbit around target)

Mission Summary: The United States' first attempt to fly to the Moon ended just seventy-seven seconds after liftoff when the rocket's first stage exploded.

Luna 1958A

Launch Date: September 23, 1958

Sponsoring Country: USSR

Type of Mission: Impactor (designed to crash into its target)

Mission Summary: The rocket carrying the spacecraft exploded a little over a minute and a half after liftoff.

Pioneer 1

Launch Date: October 11, 1958

Sponsoring Country: USA

Type of Mission: Orbiter

Mission Summary: Just after launch, the rocket's second and third stages failed to separate evenly. As a result, the spacecraft was unable to achieve a lunar trajectory (a path that would take it to the Moon). However, *Pioneer 1* did manage to return data on the Van Allen Belt and other phenomena before reentering Earth's atmosphere on October 12, 1958.

Luna 1958B

Launch Date: October 11, 1958

Sponsoring Country: USSR

Type of Mission: Impactor

Mission Summary: The rocket carrying the spacecraft to the Moon exploded just after liftoff. Although *Luna 1958B* was launched hours after *Pioneer 1*, its trajectory would have allowed it to beat *Pioneer 1* to the Moon.

Pioneer 2

Launch Date: November 8, 1958

Sponsoring Country: USA

Type of Mission: Orbiter

Mission Summary: Just after launch, the third stage of the rocket failed to ignite. As a result, the spacecraft did not achieve orbit and fell back to Earth.

Luna 1958C

Launch Date: December 4, 1958

Sponsoring Country: USSR

Type of Mission: Impactor

Mission Summary: The rocket's first stage failed about three minutes after launch and the spacecraft fell back to Earth.

Pioneer 3

Launch Date: December 6, 1958

Sponsoring Country: USA

Type of Mission: Flyby of Moon

Mission Summary: The rocket's first stage shut off early, causing the spacecraft to crash.

Luna 1

Launch Date: January 2, 1959

Sponsoring Country: USSR

Type of Mission: Impactor

Mission Summary: During its flight to the Moon, a rocket used to carry out a trajectory correction maneuver failed. As a result, *Luna 1* missed the Moon by 3,723 miles (5,995 kilometers) and went into a solar orbit.

Pioneer 4—the first spacecraft to fly by the Moon

Launch Date: March 3, 1959

Arrival at Moon: March 4, 1959

Sponsoring Country: USA

Type of Mission: Flyby of Moon

Mission Summary: *Pioneer 4* passed within 37,300 miles (60,000 kilometers) of the Moon and returned data on lunar radiation levels. The spacecraft then continued on and entered into a solar orbit.

Luna 1959A

Launch Date: June 18, 1959

Sponsoring Country: USSR

Type of Mission: Impactor

Mission Summary: The rocket's guidance system failed about two and a half minutes into flight and the spacecraft was unable to reach Earth orbit.

Luna 2—the first spacecraft to land on another celestial body

Launch Date: September 12, 1959

Arrival at Moon: September 14, 1959

Sponsoring Country: USSR

Type of Mission: Impactor

Mission Summary: *Luna 2* was the first spacecraft to land on another celestial body. The spacecraft impacted on the Moon's surface at Palus Putredinis, just east of the Sea of Serenity near the craters Aristides, Archimedes, and Autolycus.

Luna 3—the mission that sent back the first images of the far side of the Moon

Launch Date: October 4, 1959

Arrival at Moon: October 6, 1959

Sponsoring Country: USSR

Type of Mission: Flyby of Moon and return

Mission Summary: *Luna 3* was the first spacecraft to send back pictures of the far side of the Moon. The spacecraft took twenty-nine images over a forty-minute period as it flew past the lunar surface. Three images were released to the public, as well as a composite image of the entire far side. With these pictures, the Soviets were the first to name features on the Moon's far side. *Luna 3*'s trajectory took the spacecraft from Earth, around the Moon and back, where it reentered Earth's atmosphere on April 20, 1960.

Pioneer P–3

Launch Date: November 26, 1959

Sponsoring Country: USA

Type of Mission: Flyby of Moon

Mission Summary: The protective cover that surrounded the spacecraft while it was attached to the rocket broke away after only forty-five seconds in flight. As a result, the spacecraft failed to reach orbit and crashed back to Earth.

Luna 1960A

Launch Date: April 15, 1960

Sponsoring Country: USSR

Type of Mission: Flyby of Moon

Mission Summary: The rocket's second stage shut off early and the spacecraft was unable to reach Earth orbit. It fell back to Earth and was destroyed.

Luna 1960B

Launch Date: April 19, 1960

Sponsoring Country: USSR

Type of Mission: Flyby of Moon

Mission Summary: A malfunction caused the rocket carrying the spacecraft to explode just after liftoff, damaging the launch pad in the process.

Pioneer P–30

Launch Date: September 25, 1960

Sponsoring Country: USA

Type of Mission: Orbiter

Mission Summary: The rocket's second stage failed and the spacecraft was unable to reach Earth orbit.

Pioneer P–31

Launch Date: December 15, 1960

Sponsoring Country: USA

Type of Mission: Orbiter

Mission Summary: The rocket's first stage malfunctioned just one minute after launch and the spacecraft crashed into the Atlantic Ocean.

Ranger 1

Launch Date: August 23, 1961

Sponsoring Country: USA

Type of Mission: Test of lunar parking orbit

Mission Summary: *Ranger 1* was designed to test the feasibility of going into a parking orbit around Earth before heading out to the Moon. A parking orbit is simply a temporary orbit that gives engineers a little extra time to calculate a more accurate trajectory to the Moon based on data received from the launch. *Ranger 1* successfully made it into low Earth orbit. Then, thirteen minutes later when its engines reignited to give the spacecraft an additional boost, they burned only for a few seconds and then shut off. Without the full ninety-second burn that was planned, the spacecraft was unable to leave Earth orbit and eventually reentered the atmosphere after completing 111 orbits.

Ranger 2

Launch Date: November 18, 1961

Sponsoring Country: USA

Type of Mission: Test of lunar parking orbit

Mission Summary: *Ranger 2*, like *Ranger 1*, was designed as a test vehicle. The spacecraft's engines failed to reignite after the spacecraft entered into a low Earth orbit. As a result, *Ranger 2* burned up in Earth's atmosphere just two days after launch.

Ranger 3

Launch Date: January 26, 1962

Sponsoring Country: USA

Type of Mission: Lander (designed to land on its target)

Mission Summary: Designed to take close-up images of the Moon before impacting with its surface, *Ranger 3* missed the Moon and ended up in a solar orbit.

Ranger 4

Launch Date: April 23, 1962

Arrival at Moon: April 26, 1962

Sponsoring Country: USA

Type of Mission: Lander

Mission Summary: After a successful launch, a failure of some sort onboard *Ranger 4* made communication with the spacecraft impossible. Engineers were able to track the spacecraft until it crashed into the far side of the Moon, but they were unable to collect any data.

Ranger 5

Launch Date: October 18, 1962

Sponsoring Country: USA

Type of Mission: Lander

Mission Summary: Solar cells onboard *Ranger 5* failed shortly after launch. Without power, engineers on the ground were unable to control the spacecraft and *Ranger 5* missed the Moon by 450 miles (720 kilometers).

Sputnik 25

Launch Date: January 4, 1963

Sponsoring Country: USSR

Type of Mission: Lander

Mission Summary: After a successful launch, the spacecraft failed to transfer to a lunar trajectory. *Sputnik 25* burned up as it reentered Earth's atmosphere.

Luna 1963B

Launch Date: February 2, 1963

Sponsoring Country: USSR

Type of Mission: Lander

Mission Summary: Ground controllers lost control of the rocket almost five minutes after liftoff and the spacecraft failed to reach Earth orbit.

Luna 4

Launch Date: April 2, 1963

Arrival at Moon: April 6, 1963

Sponsoring Country: USSR

Type of Mission: Orbiter

Mission Summary: After a successful launch, a rocket used for a trajectory correction maneuver failed to operate and *Luna 4* was unable to adjust its course to go into lunar orbit. Contact with *Luna 4* was lost after it passed within 5,780 miles (9,300 kilometers) of the Moon.

Ranger 6

Launch Date: January 30, 1964

Arrival at Moon: February 2, 1964

Sponsoring Country: USA

Type of Mission: Impactor

Mission Summary: *Ranger 6* was designed to take a series of images as it approached the

Moon, right up to the point where it crashed into its surface. Unfortunately, the spacecraft's cameras failed and no pictures were returned. *Ranger 6* crash-landed in the Sea of Tranquility.

Luna 1964A

Launch Date: March 21, 1964

Sponsoring Country: USSR

Type of Mission: Lander

Mission Summary: Problems with the rocket's third stage prevented the spacecraft from reaching orbit. The spacecraft fell back to Earth.

Luna 1964B

Launch Date: April 20, 1964

Sponsoring Country: USSR

Type of Mission: Lander

Mission Summary: The rocket carrying the spacecraft failed. As a result, the spacecraft was unable to reach Earth orbit.

Zond 1964A

Launch Date: June 4, 1964

Sponsoring Country: USSR

Type of Mission: Flyby of Moon

Mission Summary: The spacecraft failed to reach Earth orbit.

Ranger 7

Launch Date: July 28, 1964

Arrival at Moon: July 31, 1964

Sponsoring Country: USA

Type of Mission: Impactor

Mission Summary: *Ranger 7* sent back the first high-quality images of the lunar surface before it crash-landed in the Known Sea (Mare Cognitum). A total of more than 4,300 images were sent back.

Ranger 8

Launch Date: February 17, 1965

Arrival at Moon: February 20, 1965

Sponsoring Country: USA

Type of Mission: Impactor

Mission Summary: *Ranger 8* took more than 7,100 high-quality images of the lunar surface before it crash-landed in the Sea of Tranquility.

Cosmos 60

Launch Date: March 12, 1965

Sponsoring Country: USSR

Type of Mission: Lander

Mission Summary: The spacecraft made it into Earth orbit but for some reason was unable to continue on to the Moon. It is possible that the spacecraft is still in orbit around Earth today.

Ranger 9

Launch Date: March 21, 1965

Arrival at Moon: March 24, 1965

Sponsoring Country: USA

Type of Mission: Impactor

Mission Summary: *Ranger 9* took more than 5,800 images of the lunar surface before it crash-landed in the crater Alphonsus. Network television broadcasted images from the spacecraft as they were received—live from the Moon!

Luna 1965A

Launch Date: April 10, 1965

Sponsoring Country: USSR

Type of Mission: Lander

Mission Summary: Failure of the rocket's third stage kept the spacecraft from reaching orbit.

Luna 5

Launch Date: May 9, 1965

Arrival at Moon: May 12, 1965

Sponsoring Country: USSR

Type of Mission: Lander

Mission Summary: This mission was the first-ever attempt to "soft-land" a spacecraft on the Moon. *Luna 5*'s retro-rockets failed to fire and the spacecraft crash-landed near the Sea of Clouds.

Luna 6

Launch Date: June 8, 1965

Sponsoring Country: USSR

Type of Mission: Lander

Mission Summary: On its way to the Moon, a rocket failed to turn off after a trajectory correction maneuver. As a result, *Luna 8* missed the Moon and went into a solar orbit.

Zond 3

Launch Date: July 18, 1965

Arrival at Moon: July 20, 1965

Sponsoring Country: USSR

Type of Mission: Flyby of Moon

Mission Summary: *Zond 3* took twenty-five images as it flew within 5,720 miles (9,200 kilometers) of the far side of the Moon. It transmitted the images back to Earth nine days later. After passing the Moon, the spacecraft went into a solar orbit.

Luna 7

Launch Date: October 4, 1965

Arrival at Moon: October 7, 1965

Sponsoring Country: USSR

Type of Mission: Lander

Mission Summary: *Luna 7* made it to the Moon, but its retro-rockets switched on too soon. The spacecraft crash-landed in the Ocean of Storms, west of the crater Kepler.

Luna 8

Launch Date: December 3, 1965

Arrival at Moon: December 6, 1965

Sponsoring Country: USSR

Type of Mission: Lander

Mission Summary: *Luna 8* made it to the Moon, but its retro-rockets fired too late and the spacecraft crash-landed in the Ocean of Storms, east of the crater Galilaei.

Luna 9—the first controlled landing on the Moon

Launch Date: January 31, 1966

Arrival at Moon: February 3, 1966

Sponsoring Country: USSR

Type of Mission: Lander

Mission Summary: *Luna 9* was the first spacecraft to make a controlled landing onto the surface of another celestial body. Scientists believe the spacecraft landed on the sloping floor of a shallow crater. Over the next two days, the spacecraft sent back three panoramas of the lunar landscape. Between the second and third transmissions, the spacecraft evidently shifted or settled a few centimeters, because the third batch of images was taken from a slightly different angle. The different angle allowed scientists to construct a stereoview of the landing site and determine the distances to various rocks and depressions. The last communication with the spacecraft was on February 5, 1966.

Cosmos 111

Launch Date: March 1, 1966

Sponsoring Country: USSR

Type of Mission: Flyby of Moon

Mission Summary: The spacecraft was unable to achieve a lunar trajectory. It was destroyed as it reentered Earth's atmosphere on March 3, 1966.

Luna 10—the first spacecraft to go into orbit around another celestial body

Launch Date: March 31, 1966

Arrival at Moon: April 2, 1966

Sponsoring Country: USSR

Type of Mission: Orbiter

Mission Summary: When *Luna 10* arrived at the Moon, it became the first spacecraft to successfully go into orbit around another celestial body. While in lunar orbit, *Luna 10* took readings of radiation levels, cosmic ray intensities, and the Moon's weak magnetic field. The spacecraft successfully transmitted data for two months, circling the Moon 460 times before its mission came to an end on May 30, 1966. Although all communication was terminated, *Luna 10* continued in orbit around the Moon. By now, the uneven tug of the Moon's gravity has probably pulled the spacecraft from orbit, crashing it into the lunar surface.

Luna 1966A

Launch Date: April 30, 1966

Sponsoring Country: USSR

Type of Mission: Orbiter

Mission Summary: The spacecraft failed to achieve orbit and came crashing back to Earth.

Surveyor 1

Launch Date: May 30, 1966

Arrival at Moon: June 2, 1966

Sponsoring Country: USA

Type of Mission: Lander

Mission Summary: *Surveyor 1* was the first spacecraft from the United States to perform a controlled landing on the surface of the Moon. After touchdown, *Surveyor 1* took more than 11,100 images of the lunar landscape. Its mission lasted six weeks.

Explorer 33

Launch Date: July 1, 1966

Sponsoring Country: USA

Type of Mission: Orbiter

Mission Summary: The spacecraft was designed to study interplanetary space while in lunar orbit. Even though *Explorer 33* failed to go into orbit around the Moon, the elliptical orbit it ended up in allowed the spacecraft to complete most of its mission. Contact with the spacecraft was lost around September 21, 1971.

Lunar Orbiter 1

Launch Date: August 10, 1966

Arrival at Moon: August 14, 1966

Sponsoring Country: USA

Type of Mission: Orbiter

Mission Summary: *Lunar Orbiter 1* went into orbit around the Moon and sent back high-quality images (by television) of more than two million square miles of lunar surface, including the first detailed images of potential Apollo landing sites. After circling the Moon 527 times in seventy-seven days, engineers on Earth deliberately crashed the spacecraft onto the Moon's surface, so that it wouldn't interfere with the upcoming manned missions.

Luna 11

Launch Date: August 24, 1966

Arrival at Moon: August 27, 1966

Sponsoring Country: USSR

Type of Mission: Orbiter

Mission Summary: Upon its arrival, *Luna 11* successfully went into a lunar orbit. The spacecraft, designed to test new technology, completed 277 orbits before its mission was terminated on October 1, 1966. Although all communications with the spacecraft ended, *Luna*

11 continued to orbit the Moon. By now, the uneven tug of the Moon's gravity has probably pulled the spacecraft from orbit, crashing it into the lunar surface.

Surveyor 2

Launch Date: September 20, 1966

Arrival at Moon: September 22, 1966

Sponsoring Country: USA

Type of Mission: Lander

Mission Summary: Just before touching down on the lunar surface, one of the spacecraft's thrusters malfunctioned during a midcourse correction and *Surveyor 2* tumbled out of control. It crashed into the Moon, southeast of the crater Copernicus.

Luna 12

Launch Date: October 22, 1966

Arrival at Moon: October 25, 1966

Sponsoring Country: USSR

Type of Mission: Orbiter

Mission Summary: When *Luna 12* went into lunar orbit, its primary mission was to photograph the lunar surface and it did just that. The spacecraft took 1,100 pictures, including images of the Sea of Rains and the area surrounding the crater Aristarchus. The mission was terminated on January 19, 1967, after 602 orbits, although the spacecraft continued in orbit for some time. By now, the Moon's uneven gravity has probably pulled the spacecraft from orbit, crashing it into the lunar surface.

Lunar Orbiter 2

Launch Date: November 6, 1966

Arrival at Moon: November 10, 1966

Sponsoring Country: USA

Type of Mission: Orbiter

Mission Summary: *Lunar Orbiter 2* went into lunar orbit and took more than eight hundred pictures during its mission, including an oblique view of the crater Copernicus that was voted one of the best images of the century by the press. (Images of the Moon improved dramatically with later missions.) After successfully completing its mission, the spacecraft was deliberately sent crashing into the lunar surface on October 11, 1967. Crashing the spacecraft into the Moon kept the space around it clear of debris that might compromise the upcoming manned missions.

Luna 13

Launch Date: December 21, 1966

Arrival at Moon: December 24, 1966

Sponsoring Country: USSR

Type of Mission: Lander

Mission Summary: *Luna 13* bounced to a landing on the lunar surface, coming to a rest in the Ocean of Storms between the craters Selencus and Craft. The lander collected soil samples and conducted experiments to determine the soil density and radioactivity. The mission ended on December 30, 1966, when the spacecraft's supplies were depleted.

Lunar Orbiter 3

Launch Date: February 4, 1967

Arrival at Moon: February 8, 1967

Sponsoring Country: USA

Type of Mission: Orbiter

Mission Summary: Once *Lunar Orbiter 3* was safely installed into a lunar orbit, engineers on Earth began experimenting with the spacecraft. By altering its orbit several times over the course of its mission, controllers on Earth gained experience communicating with a spacecraft in various orientations around the Moon. *Lunar Orbiter 3* was also able to photograph *Surveyor 2* on the surface. The mission ended on October 9, 1967, when controllers deliberately crashed the spacecraft into the Moon.

Surveyor 3

Launch Date: April 17, 1967

Arrival at Moon: April 20, 1967

Sponsoring Country: USA

Type of Mission: Lander

Mission Summary: As *Surveyor 3* came in for a soft landing on the Moon, one of its thrusters didn't turn off at the proper time and the spacecraft bounced a couple of times before it came to rest in the Ocean of Storms. Despite the rough landing, the spacecraft survived. During its mission, the lander used a scoop to collect soil samples and its camera took more than 6,300 images.

Lunar Orbiter 4

Launch Date: May 4, 1967

Arrival at Moon: May 8, 1967

Sponsoring Country: USA

Type of Mission: Orbiter

Mission Summary: Soon after *Lunar Orbiter 4* went into orbit around the Moon, it was able to take the first pictures of the Moon's south pole. It took images from orbit for eight months before controllers sent the spacecraft crashing to the lunar surface, again in an effort to keep the lunar space free of debris.

Surveyor 4

Launch Date: July 14, 1967

Arrival at Moon: July 17, 1967

Sponsoring Country: USA

Type of Mission: Lander

Mission Summary: Controllers lost contact with *Surveyor 4* just two and a half minutes before it was to touch down in Sinus Medii.

Explorer 35 (IMP-E)

Launch Date: July 19, 1967

Sponsoring Country: USA

Type of Mission: Orbiter

Mission Summary: After launch, *Explorer 35* was placed in a highly elliptical orbit that took it around the Moon. The spacecraft successfully studied interplanetary space for six years before its mission was terminated on June 24, 1973.

Lunar Orbiter 5

Launch Date: August 1, 1967

Arrival at Moon: August 5, 1967

Sponsoring Country: USA

Type of Mission: Orbiter

Mission Summary: Upon completion of *Lunar Orbiter 5*'s mission, more than 99 percent of the Moon's surface had been mapped (with data from all previous missions combined). The mission ended when controllers sent the spacecraft crashing to the lunar surface on January 31, 1968.

Surveyor 5

Launch Date: September 8, 1967

Arrival at Moon: September 10, 1967

Sponsoring Country: USA

Type of Mission: Lander

Mission Summary: Despite a serious helium leak that occurred during its trip to the Moon, controllers were able to bring *Surveyor 5* to a safe landing on the lunar surface. Once on the ground, controllers ordered the spacecraft to fire its engine to test the composition of the soil beneath the lander. The test firing blew away a few clumps of soil but did not create a crater. This test demonstrated to scientists that the lunar surface should be strong enough to support an Apollo lander. The final transmission from the spacecraft was received on December 17, 1967.

Zond 1967A

Launch Date: September 28, 1967

Sponsoring Country: USSR

Type of Mission: Unmanned test of lunar capsule

Mission Summary: One of the rocket engines shut down early causing it to veer off course. The capsule was safely jettisoned, while the rocket crashed.

Surveyor 6—the first and only intentional liftoff and landing of a spacecraft on the Moon

Launch Date: November 7, 1967

Arrival at Moon: November 9, 1967

Sponsoring Country: USA

Type of Mission: Lander

Mission Summary: *Surveyor* 6 touched down on the Moon in Sinus Medii where it began taking a series of pictures and soil samples. Then, on November 17, controllers ordered the spacecraft's engines to fire, lifting the lander off the lunar surface ten feet (three meters) and setting it down again a few feet from its original landing site. *Surveyor* 6 then took pictures of the former landing site, checking for evidence of a crater created by the rocket's exhaust. No crater was found, indicating that the Moon's surface was solid. The last contact with the spacecraft was on December 14, 1967.

Zond 1967B

Launch Date: November 22, 1967

Sponsoring Country: USSR

Type of Mission: Unmanned test of lunar capsule

Mission Summary: The rocket's second stage failed and the spacecraft fell back to Earth.

Surveyor 7

Launch Date: January 7, 1968

Arrival at Moon: January 9, 1968

Sponsoring Country: USA

Type of Mission: Lander

Mission Summary: *Surveyor 7* landed in the lunar highlands near the crater Tycho. Scientists used the scoop on the spacecraft to "weigh" lunar rocks, based on how much current was needed to lift each rock. Images sent back from the spacecraft indicated, for the first time, that some of the lunar rocks had been molten at some time in their history. The mission was successfully completed on February 21, 1968.

Luna 1968A

Launch Date: February 7, 1968
Sponsoring Country: USSR
Type of Mission: Orbiter
Mission Summary: The spacecraft failed to reach orbit and fell back to Earth.

Luna 14

Launch Date: April 7, 1968
Arrival at Moon: April 10, 1968
Sponsoring Country: USSR
Type of Mission: Orbiter
Mission Summary: *Luna 14* went into orbit around the Moon and began taking pictures of the lunar surface and studying the lunar gravitational field. Controllers also used this mission to test their communication and tracking abilities in preparation for manned lunar missions. Once its mission was complete, *Luna 14* continued in orbit around the Moon. By now, however, the uneven tug of the Moon's gravity has probably pulled the spacecraft from orbit, causing it to crash into the lunar surface.

Zond 1968A

Launch Date: April 23, 1968
Sponsoring Country: USSR
Type of Mission: Test flight of lunar systems
Mission Summary: The rocket's second stage shut down prematurely and the spacecraft crashed back to Earth.

Zond 5—the first turtles to fly to the Moon and back

Launch Date: September 15, 1968
Arrival at Moon: September 18, 1968
Sponsoring Country: USSR
Type of Mission: Flyby of Moon and return
Mission Summary: *Zond 5* flew within 1,211 miles (1,950 kilometers) of the lunar surface and then returned to Earth, splashing down in the Indian Ocean on September 21, 1968.

The spacecraft was recovered and taken back to the USSR for study. Within the payload compartment of the spacecraft, scientists had placed live turtles, flies, worms, seeds, and other biological specimens. After their flight, the turtles had lost some weight but otherwise seemed no worse for their trip to the Moon. There is little information about the status of the other creatures onboard. Many saw *Zond 5* as one of the last steps before the Soviet Union landed cosmonauts on the Moon.

Zond 6

Launch Date: November 10, 1968

Arrival at Moon: November 14, 1968

Sponsoring Country: USSR

Type of Mission: Flyby of Moon and return

Mission Summary: *Zond 6* was seen by the Western powers as being the Soviet Union's final test before launching cosmonauts to the Moon. Once the spacecraft left Earth orbit, it took two days to reach the Moon. There, it took pictures as it swung close to the lunar surface. *Zond 6* then returned to Earth. Instead of splashing down in the Indian Ocean, like the previous *Zond 5* mission, controllers programmed the spacecraft to bounce off the atmosphere and redirected the capsule to parachute to a landing within Soviet territory on November 17, 1968.

Apollo 8—the first mission to take humans beyond the gravitational pull of Earth

Launch Date: December 21, 1968

Arrival at Moon: December 24, 1968

Return to Earth: December 27, 1968

Sponsoring Country: USA

Type of Mission: Orbiter

Mission Summary: Astronauts Frank Borman, James Lovell, and William Anders were the first humans to leave the Earth and travel to the Moon. They arrived at the Moon on December 24, 1968, completed ten orbits, and returned to Earth.

Zond 1969A

Launch Date: January 20, 1969

Sponsoring Country: USSR

Type of Mission: Flyby of Moon and return

Mission Summary: The spacecraft's second stage rocket shut down early and the rocket failed to achieve Earth orbit.

Luna 1969A

Launch Date: February 19, 1969

Sponsoring Country: USSR

Type of Mission: Rover

Mission Summary: The spacecraft's rocket exploded shortly after launch.

Zond L1S–1

Launch Date: February 21, 1969

Sponsoring Country: USSR

Type of Mission: Orbiter

Mission Summary: The spacecraft's rocket exploded shortly after launch.

Luna 1969B

Launch Date: April 15, 1969

Sponsoring Country: USSR

Type of Mission: Sample return (designed to bring a sample of its target back to Earth)

Mission Summary: The spacecraft's rocket apparently exploded on the launch pad.

Apollo 10

Launch Date: May 18, 1969

Arrival at Moon: May 21, 1969

Return to Earth: May 26, 1969

Sponsoring Country: USA

Type of Mission: Orbiter

Mission Summary: Astronauts Thomas Stafford, John Young, and Eugene Cernan went into lunar orbit on May 21, 1969, where they tested procedures for the first Moon landing. *Apollo 10* splashed down in the Pacific Ocean.

Luna 1969C

Launch Date: June 14, 1969

Sponsoring Country: USSR

Type of Mission: Sample return

Mission Summary: The spacecraft's rocket exploded shortly after launch.

Zond L1S–2

Launch Date: July 3, 1969

Sponsoring Country: USSR

Type of Mission: Orbiter

Mission Summary: The rocket's engines shut down just moments after launch, causing the spacecraft to come crashing back to Earth.

Luna 15

Launch Date: July 13, 1969
Arrival at Moon: July 17, 1969
Sponsoring Country: USSR
Type of Mission: Sample return
Mission Summary: *Luna 15* was launched in a veil of secrecy only three days before *Apollo 11*. The Soviets did not reveal the target or mission of *Luna 15*, causing some concern on the part of the United States. Would *Luna 15* interfere with *Apollo 11*? Where would it land? Would there be communication interference? Just hours before the arrival of *Apollo 11*, *Luna 15* began its descent to the lunar surface in the Sea of Crises. However, the spacecraft appeared to crash-land. The Soviets issued a statement claiming the research part of the mission was complete, but there had been hints that *Luna 15* was intended to be a sample return mission. No samples were returned.

Apollo 11—the first humans on the Moon

Launch Date: July 16, 1969
Arrival at Moon: July 20, 1969
Return to Earth: July 24, 1969
Sponsoring Country: USA
Type of Mission: First manned Moon landing
Mission Summary: While astronaut Michael Collins orbited overhead, Neil Armstrong and Edwin "Buzz" Aldrin became the first men to land on the Moon. They landed in the Sea of Tranquility on July 20, 1969. They remained on the lunar surface twenty-one and a half hours and collected forty-six pounds (twenty-one kilograms) of samples. When *Apollo 11* returned to Earth, the capsule splashed down in the Pacific Ocean to a hero's welcome.

Zond 7

Launch Date: August 7 (or 8), 1969
Arrival at Moon: August 11, 1969
Sponsoring Country: USSR
Type of Mission: Flyby of Moon and return
Mission Summary: *Zond 7* flew to the Moon and back, taking color pictures of the Earth and lunar surface along the way. The spacecraft soft-landed back on Earth in Kazakstan on August 14, 1969.

Cosmos 300

Launch Date: September 23, 1969

Sponsoring Country: USSR

Type of Mission: Sample return

Mission Summary: The spacecraft reached Earth orbit, but then its rocket failed and the spacecraft was unable to continue on to the Moon.

Cosmos 305

Launch Date: October 22, 1969

Sponsoring Country: USSR

Type of Mission: Sample return

Mission Summary: The spacecraft's rocket failed once it reached Earth orbit. As a result, it was unable to continue on to the Moon.

Apollo 12

Launch Date: November 14, 1969

Arrival at Moon: November 19, 1969

Return to Earth: November 24, 1969

Sponsoring Country: USA

Type of Mission: Manned Moon landing

Mission Summary: *Apollo 12* began its mission with an extra jolt, literally. Just seconds after liftoff, *Apollo 12*'s *Saturn V* rocket was struck by lightning. After a few tense moments, ground controllers reset some circuit breakers and everything showed that the spacecraft was still on course. The mission was allowed to continue on its way. Once at the Moon, astronaut Richard Gordon orbited overhead while Charles (Pete) Conrad and Alan Bean landed on the lunar surface on November 19, 1969. *Apollo 12* touched down in the Ocean of Storms, within walking distance of the *Surveyor 3* spacecraft. The astronauts were on the lunar surface for thirty-one and a half hours and collected seventy-five pounds (thirty-four kilograms) of samples.

Luna 1970A

Launch Date: February 6, 1970

Sponsoring Country: USSR

Type of Mission: Sample return

Mission Summary: The rocket failed to reach orbit and the spacecraft fell back to Earth.

Luna 1970B

Launch Date: February 19, 1970

Sponsoring Country: USSR

Type of Mission: Orbiter

Mission Summary: The spacecraft failed to reach Earth orbit.

Apollo 13

Launch Date: April 11, 1970

Arrival at Moon: April 15, 1970

Return to Earth: April 17, 1970

Sponsoring Country: USA

Type of Mission: Manned Moon landing

Mission Summary: When *Apollo 13* was halfway to the Moon, an explosion in the spacecraft's Service Module required mission control to cancel the scheduled Moon landing and focus on bringing astronauts James Lovell, Fred Haise, and John Swigert safely home. The mission was called a successful failure. While they failed to land on the Moon, the men did return home safely.

Luna 16—the first successful robotic sample return mission

Launch Date: September 12, 1970

Arrival at Moon: September 17, 1970

Sponsoring Country: USSR

Type of Mission: Sample return

Mission Summary: *Luna 16* was the first robotic mission to land on the Moon, collect samples of dust and rock, and return those samples to Earth. *Luna 16* was also the first spacecraft to land in the lunar darkness. The spacecraft landed on September 20, 1970, in the Sea of Fertility. Twenty-six hours later, after collecting dust and rock samples, the spacecraft was launched back into space. It returned to Earth with a soft landing on September 24, 1970, bringing back 101 grams of Moon rocks.

Zond 8

Launch Date: October 20, 1970

Arrival at Moon: October 24, 1970

Sponsoring Country: USSR

Type of Mission: Flyby of Moon and return

Mission Summary: *Zond 8* flew by the Moon, taking pictures and gathering data. Then the spacecraft returned to Earth, splashing down in the Indian Ocean on October 27, 1970.

Luna 17 (Lunokhod 1)—the first remote-controlled rover on the Moon

Launch Date: November 10, 1970

Arrival at Moon: November 15, 1970

Sponsoring Country: USSR

Type of Mission: Rover

Mission Summary: *Luna 17* gently touched down on the lunar surface in the Sea of Rains. Two ramps extended away from the spacecraft, allowing the Lunokhod rover to roll onto the lunar soil. Over the course of eleven lunar cycles, the remote-controlled rover traveled more than six miles (ten and a half kilometers) and sent back twenty thousand television pictures of the lunar landscape. The rover was officially shut down on October 4, 1971.

Apollo 14—the first golf swing on the Moon

Launch Date: January 31, 1971

Arrival at Moon: February 5, 1971

Return to Earth: February 9, 1971

Sponsoring Country: USA

Type of Mission: Manned Moon landing

Mission Summary: While astronaut Stuart Roosa orbited overhead, Alan Shepard (the first American in space) and Edgar Mitchell landed on the Moon on February 5, 1971. *Apollo 14* touched down in the Fra Mauro highlands, the original landing for *Apollo 13*. While exploring the lunar surface, Alan Shepard took time out of his busy day to play a bit of golf. Using a golf club he constructed out of various tools and two golf balls he had managed to sneak onboard, Shepard became the first human to play golf on another world. The astronauts were on the lunar surface for thirty-three and a half hours and collected ninety-five pounds (forty-three kilograms) of samples. They returned to Earth on February 9, 1971.

Apollo 15—the first manned mission to carry a rover

Launch Date: July 26, 1971

Arrival at Moon: July 30, 1971

Return to Earth: August 7, 1971

Sponsoring Country: USA

Type of Mission: Manned Moon landing

Mission Summary: While astronaut Alfred Worden orbited overhead, David Scott and James Irwin landed on the Moon on July 30, 1971, in the Hadley Rille region. *Apollo 15* was the first lander to carry a lunar rover, allowing astronauts to explore a much larger area of the Moon's surface. Scott and Worden drove the rover almost seventeen miles (twenty-eight kilometers). They were on the lunar surface for almost sixty-seven hours and collected 170 pounds (77 kilograms) of samples.

Luna 18

Launch Date: September 2, 1971

Arrival at Moon: September 7, 1971

Sponsoring Country: USSR

Type of Mission: Lander

Mission Summary: Although its primary mission was a lander, *Luna 18* first went into orbit upon its arrival at the Moon. The spacecraft completed fifty-four orbits before firing its braking thrusters and beginning its descent to the lunar surface. The mission failed, however, when the spacecraft impacted the surface on September 11, 1971, and ceased communicating with controllers on Earth.

Luna 19

Launch Date: September 28, 1971

Arrival at Moon: October 3, 1971

Sponsoring Country: USSR

Type of Mission: Orbiter

Mission Summary: *Luna 19* went into orbit around the Moon where it studied radiation levels, mass concentrations, and other aspects of the lunar environment. The spacecraft also studied the solar wind. Once its mission was complete, *Luna 19* continued in orbit around the Moon. By now, however, the uneven tug of the Moon's gravity has probably pulled the spacecraft from orbit, causing it to crash into the lunar surface.

Luna 20—the second robotic lunar sample return mission

Launch Date: February 14, 1972

Arrival at Moon: February 21, 1972

Sponsoring Country: USSR

Type of Mission: Sample return

Mission Summary: *Luna 20* soft-landed in the Apollonius highlands near the Sea of Fertility. The spacecraft collected samples and then lifted off the next day. The sealed capsule (containing thirty grams of lunar rocks and dust) landed in the Soviet Union on February 25, 1972, and was retrieved the following day.

Apollo 16—the only mission to land in the lunar highlands.

Launch Date: April 16, 1972

Arrival at Moon: April 21, 1972

Return to Earth: April 27, 1972

Sponsoring Country: USA

Type of Mission: Manned Moon landing

Mission Summary: While astronaut Thomas Mattingly orbited overhead, John Young and Charles Duke landed on the Moon on April 21, 1972, in the Descartes region. *Apollo 16* carried a lunar rover that astronauts drove sixteen miles (twenty-seven kilometers). They were on the lunar surface for a little over seventy-one hours and collected 207 pounds (94 kilograms) of samples, including the oldest sample ever collected.

Soyuz L3

Launch Date: November 23, 1972

Sponsoring Country: USSR

Type of Mission: Orbiter and test vehicle

Mission Summary: *Soyuz L3* was designed to test a soyez capsule that was to function as the base for a lunar lander. Ninety seconds after launch, six of the thirty engines shut down, triggering a catastrophic failure of the launch vehicle.

Apollo 17—the last manned mission to the moon

Launch Date: December 7, 1972

Arrival at Moon: December 11, 1972

Return to Earth: December 19, 1972

Sponsoring Country: USA

Type of Mission: Manned Moon landing

Mission Summary: While astronaut Ronald Evans orbited overhead, Eugene Cernan and Harrison (Jack) Schmitt landed in the Taurus-Littrow region on the Moon on December 11, 1972. Schmitt was the first geologist to land on the Moon. Cernan and Schmitt drove eighteen miles (thirty kilometers) in their lunar rover, collected 242 pounds (110 kilograms) of samples, and spent seventy-five hours on the lunar surface. *Apollo 17* returned to Earth and brought to a close humans' exploration of the Moon—to date.

Luna 21 (Lunokhod 2)—the second remote-controlled lunar rover mission

Launch Date: January 8, 1973

Arrival at Moon: January 15, 1973

Sponsoring Country: USSR

Type of Mission: Rover

Mission Summary: When *Luna 21* first landed on the Moon, its rover began by taking a panoramic shot of the landing site. Then, *Lunokhod 2* rolled off of its protective shell and onto the lunar surface. The rover was powered by solar panels and kept warm at night by a radioactive heat source. The mission lasted four months, during which it took more than eighty thousand TV pictures and covered twenty-two miles (thirty-seven kilometers) of lunar terrain.

Explorer 49 (RAE-B)

Launch Date: June 10, 1973

Arrival at Moon: June 15, 1973

Sponsoring Country: USA

Type of Mission: Orbiter

Mission Summary: While this mission went into orbit around the Moon, it was not designed to study Earth's nearest neighbor. Instead, its primary mission was one dedicated to radio astronomy. Using the Moon to block out radio interference from our home planet, astronomers were able to study the radio signatures of the other planets as well as objects far beyond our solar system. *Explorer 49*'s primary mission ended in June 1975, but occasional contact with the spacecraft was made until August of 1977. This was the last US mission to the Moon for twenty-one years.

Luna 22

Launch Date: May 29, 1974

Arrival at Moon: June 2, 1974

Sponsoring Country: USSR

Type of Mission: Orbiter

Mission Summary: *Luna 22* studied the Moon's magnetic field, gamma ray emissions, and gravitational field while in lunar orbit. Its mission ended in November of 1975 after the spacecraft used up all of its maneuvering fuel. *Luna 22* continued in orbit around the Moon. By now, however, the uneven tug of the Moon's gravity has probably pulled the spacecraft from orbit, causing it to crash into the lunar surface.

Luna 23

Launch Date: October 28, 1974

Arrival at Moon: November 6, 1974

Sponsoring Country: USSR

Type of Mission: Sample return

Mission Summary: *Luna 23* was damaged during its landing on the lunar surface. The spacecraft was thus unable to collect samples of the lunar soil. The spacecraft transmitted data for three days before falling silent.

Luna 1975A

Launch Date: October 16, 1975

Sponsoring Country: USSR

Type of Mission: Sample return

Mission Summary: The rocket's booster failed shortly after launch and the spacecraft failed to reach Earth orbit.

Luna 24

Launch Date: August 9, 1976

Arrival at Moon: August 18, 1976

Sponsoring Country: USSR

Type of Mission: Sample return

Mission Summary: After *Luna 24* touched down on the lunar soil in the Sea of Crises, it collected 170 grams of dust and rocks. The spacecraft then returned those samples to Earth on August 22, 1976.

Hiten (Muses A)—the first Japanese mission to the Moon

Launch Date: January 24, 1990

Sponsoring Country: Japan

Type of Mission: Test of lunar trajectories

Mission Summary: *Hiten* was launched into a highly elliptical Earth orbit that took it past the Moon ten times. It released *Hagoromo*, a small spacecraft that was to go into lunar orbit, but its transmitter failed before it reached the Moon. The Japanese used *Hiten* to test various technologies for future lunar missions. The spacecraft was intentionally crashed into the Moon on April 10, 1993.

Clementine

Launch Date: January 25, 1994

Arrival at Moon: February 21, 1994

Sponsoring Country: USA

Type of Mission: Orbiter

Mission Summary: *Clementine* was launched by the United States Department of Defense to demonstrate new technologies. The spacecraft flew by Earth twice during the first month of its mission before going into orbit around the Moon. Once in lunar orbit, *Clementine* began its primary seventy-day mapping mission. Then the spacecraft entered a circumlunar orbit and was to have flown on to an encounter with the asteroid Geographos in July 1994. However, a malfunctioning thruster depleted all of its maneuvering fuel, leaving the spacecraft stuck in Earth orbit. It lost power in June 1994, after studying the Van Allen radiation belts. Its remains continue to orbit Earth as space junk.

AsiaSat 3/HGS–1

Launch Date: December 24, 1997

Sponsoring Country: China

Type of Mission: Communication satellite

Mission Summary: The *AsiaSat 3* was intended to be a communication satellite for China. However, problems emerged during launch and the spacecraft was unable to reach its proper, geosynchronous orbit. It ended up in an elliptical orbit that made it unusable as a communication satellite. The mission was declared a failure, and operation of the satellite was taken over by Hughes Aircraft. After some creative thinking on the part of ground controllers, they were able to send the spacecraft flying toward the Moon. By flying past Earth's nearest neighbor twice and using the Moon's gravity to tweak its orbit, ground controllers were finally able to settle *AsiaSat 3* into its proper orbit around Earth.

Lunar Prospector

Launch Date: January 7, 1998

Arrival at Moon: January 11, 1998

Sponsoring Country: USA

Type of Mission: Orbiter

Mission Summary: *Lunar Prospector* was designed to go into a low polar orbit around the Moon and search for water ice and minerals in the dark areas of craters that get very little, if any, sunlight. During its nineteen-month mission, *Lunar Prospector* completed a map of the Moon's surface composition. On July 31, 1999, mission controllers crashed the spacecraft into a crater near the south pole. Observers from Earth watched for any signs of water vapor that might have been released during the impact, but none was seen.

SMART 1—the first European mission to the Moon

Launch Date: September 27, 2003

Arrival at Moon: November 15, 2004

Sponsoring Countries: Europe

Type of Mission: Orbiter

Mission Summary: *SMART 1* (which stands for *Small Missions for Advanced Research in Technology* 1) is Europe's first mission to the Moon and is designed to test new technologies. The spacecraft is powered by an ion drive, which takes a while to build up speed—hence the spacecraft's long time in flight. *SMART 1*'s primary mission is expected to last through August 2006.

MISSIONS TO MERCURY

Because Mercury is one of the three closest planets to Earth, it seems logical to assume that this world has already been explored by a flotilla of spacecraft. Think again. Only one spacecraft has made the journey to the innermost planet in our solar system, although there is another currently on its way. It's not that scientists are ignoring Mercury. It's just that physics makes this small planet a challenging target.

A spacecraft traveling directly from Earth to Mercury would typically be moving at very high speed with respect to Mercury when it arrives. This presents problems for flyby science or a spacecraft attempting to go into orbit around this small body. Missions to Mercury generally require long and complicated trajectories, gravity assists, several course correction maneuvers, and some creative thinking on the part of mission engineers. It's not impossible. It just takes a while, as shown by the following missions.

Mercury as seen by *Mariner 10*.
Image credit: NASA/JPL

THE MISSIONS

Mariner 10

Launch Date: November 3, 1973

Arrival at Mercury: March 29, 1974; September 21, 1974; March 16, 1975

Sponsoring Country: USA

Type of Mission: Flyby of Mercury

Mission Summary: *Mariner 10* was the first spacecraft to visit the innermost planet in our solar system. On its way to Mercury, it became the first spacecraft to use a "gravity assist" to alter its speed and course. By first flying close to Venus, it was able to use the gravity of that planet to decrease its speed and alter its course so it could fly on to Mercury. *Mariner 10* successfully flew by the closest planet to the Sun on three different occasions: March 29, 1974; September 21, 1974; and March 16, 1975. However, during each of these close approaches, the spacecraft flew past the same dark, nighttime side of Mercury. In order to get images of the sunlit side of the planet, the flight team successfully divided the mission into "incoming" images and "outgoing" images. *Mariner 10* took a total of 3,500 images of about half of the surface of the planet, revealing Mercury to be a barren, heavily cratered world.

MESSENGER

Launch Date: August 2, 2004

Arrival at Mercury: Flybys scheduled for January 2008, October 2008, and September 2009. Then the spacecraft is scheduled to go into orbit around Mercury in March 2011.

Sponsoring Country: USA

Type of Mission: Orbiter

Mission Summary: After launch, *MESSENGER* (which stands for *Me*rcury *S*urface, *S*pace *En*vironment, *Ge*ochemistry, and *R*anging) is scheduled to fly by Earth once (in July 2005) and Venus twice (in October 2006 and June 2007), using the gravity of the two planets to slow its speed and alter its course enough to allow the spacecraft to reach Mercury. Then the spacecraft will have to fly by Mercury an additional three times (in January 2008, October 2008, and September 2009) before it will have slowed enough to enter orbit. Once there, *MESSENGER*'s highly elliptical orbit will bring it within 124 miles (200 kilometers) of the surface, allowing scientists to study the planet as never before. Once in orbit, the mission is scheduled to last one Earth year.

MISSIONS TO VENUS

Venus is not only the second planet from the Sun, it is also the closest planet to Earth. While the planet is found within the inner solar system, its orbit is much closer to that of Earth's than Mercury's and therefore doesn't require the same complex gravitational gymnastics to get to the planet. In fact, the orbit of Venus almost seems to welcome visiting spacecraft from Earth. The planet itself, however, does not.

Venus is completely covered with a thick layer of clouds that blocks any view of its surface. These poisonous sulfur dioxide clouds allow the Sun's heat through, but keep it from radiating back into space. The result is an out-of-control greenhouse effect and an average surface temperature of 900°F (480°C), hot enough to melt lead. In addition, the planet's atmospheric pressure is almost nintey times greater than Earth's pressure at sea level. In other words, spacecraft flying by or in orbit around this planet aren't able to "see" through the clouds unless they use some type of radar. Any spacecraft

A radar image of Venus, taken by the *Magellan* spacecraft. Image credit: NASA/JPL

trying to land on the planet has to be able to survive the passage through corrosive clouds, crushing pressure, and searing heat.

Despite these challenges, a large number of spacecraft have been sent to this world. And, while many have failed, many have succeeded. As you read through the missions, here are some interesting milestones to look for:

- the first spacecraft to fly by Venus (*Mariner 2*)
- the first spacecraft to take readings from within the atmosphere of Venus (*Venera 4*)
- the first spacecraft to transmit data back from the surface of another planet (*Venera 7*)
- the first spacecraft to send back pictures of the surface of another planet (*Venera 9*)
- the first balloon to explore another planet (*Vega 1*)
- the first spacecraft to use aerobraking to change its orbit (*Magellan*)

THE MISSIONS

Sputnik 7
Launch Date: February 4, 1961
Sponsoring Country: USSR
Type of Mission: Probe
Mission Summary: The final stage of the rocket carrying *Sputnik 7* into orbit failed and the spacecraft was unable to achieve the necessary trajectory to carry it on to Venus.

Venera 1
Launch Date: February 12, 1961
Sponsoring Country: USSR
Type of Mission: Probe
Mission Summary: Communication with the spacecraft was lost while *Venera 1* was on its way to Venus.

Mariner 1
Launch Date: July 22, 1962
Sponsoring Country: USA
Type of Mission: Orbiter
Mission Summary: Shortly after launch, the rocket veered off course and was destroyed by ground controllers.

Sputnik 19

Launch Date: August 25, 1962

Sponsoring Country: USSR

Type of Mission: Probe

Mission Summary: The spacecraft made it into Earth orbit, but the rocket's last stage failed and *Sputnik 19* was unable to achieve its Venus trajectory. It reentered Earth's atmosphere three days later.

Mariner 2—the first spacecraft to fly by Venus

Launch Date: August 27, 1962

Arrival at Venus: December 14, 1962

Sponsoring Country: USA

Type of Mission: Flyby of Venus

Mission Summary: *Mariner 2* was the first spacecraft to successfully fly by Venus. The spacecraft came within 21,631 miles (34,833 kilometers) of the planet's surface and was able to detect ground temperatures as high as 800°F (428°C). Other instruments did not find any water vapor in the atmosphere or evidence of any magnetic field around the planet. Radio contact with *Mariner 2* was lost on January 3, 1963.

Sputnik 20

Launch Date: September 1, 1962

Sponsoring Country: USSR

Type of Mission: Probe

Mission Summary: The rocket's final stage failed and the spacecraft was unable to escape Earth orbit.

Sputnik 21

Launch Date: September 12, 1962

Sponsoring Country: USSR

Type of Mission: Probe

Mission Summary: The third stage of the rocket exploded shortly after liftoff, destroying the spacecraft.

Cosmos 21

Launch Date: November 11, 1963

Sponsoring Country: USSR

Type of Mission: Possible flyby of Venus

Mission Summary: It is unclear as to the exact purpose of this mission, although some believe it was to fly by the planet Venus. The spacecraft failed to escape Earth's gravity and came crashing back to Earth after three days.

Venera 1964A

Launch Date: February 19, 1964
Sponsoring Country: USSR
Type of Mission: Flyby of Venus
Mission Summary: The rocket carrying the spacecraft failed to reach Earth orbit.

Venera 1964B

Launch Date: March 1, 1964
Sponsoring Country: USSR
Type of Mission: Flyby of Venus
Mission Summary: The rocket carrying the spacecraft failed to reach Earth orbit.

Cosmos 27

Launch Date: March 27, 1964
Sponsoring Country: USSR
Type of Mission: Flyby of Venus
Mission Summary: The final stage of the rocket carrying the spacecraft into orbit failed and it was unable to achieve the necessary trajectory to carry it on to Venus.

Zond 1

Launch Date: April 2, 1964
Sponsoring Country: USSR
Type of Mission: Probe
Mission Summary: Communications with the spacecraft was lost while on its way to Venus.

Venera 2

Launch Date: November 12, 1965
Arrival at Venus: February 27, 1966
Sponsoring Country: USSR
Type of Mission: Flyby of Venus
Mission Summary: *Venera 2* flew within 15,000 miles (24,000 kilometers) of Venus on February 27, 1966, but communications with the spacecraft were lost just before its close approach with the planet.

Venera 3

Launch Date: November 16, 1965

Arrival at Venus: March 1, 1966

Sponsoring Country: USSR

Type of Mission: Atmospheric probe

Mission Summary: *Venera 3* arrived at the planet on March 1, 1966, but no data was returned. It is believed that Venus's thick atmosphere and crushing pressure destroyed the spacecraft on its way to the surface.

Cosmos 96

Launch Date: November 23, 1965

Sponsoring Country: USSR

Type of Mission: Lander

Mission Summary: After the spacecraft had reached orbit, an explosion damaged it so that it was unable to continue on to Venus. The damaged craft continued in orbit around Earth for sixteen days, then reentered the atmosphere on December 9, 1965.

Venera 1965A

Launch Date: November 23, 1965

Sponsoring Country: USSR

Type of Mission: Flyby of Venus

Mission Summary: The rocket launching this spacecraft malfunctioned and it was unable to achieve Earth orbit.

Venera 4—the first spacecraft to take readings from within the atmosphere of Venus

Launch Date: June 12, 1967

Arrival at Venus: October 18, 1967

Sponsoring Country: USSR

Type of Mission: Atmospheric probe

Mission Summary: Upon its arrival at Venus, *Venera 4* dropped several instruments, including a thermometer and a barometer, into the atmosphere. It then waited to receive data back from these probes before deploying a parachute and descending into the atmosphere itself. Preliminary readings seemed to indicate that the probe had taken measurements all the way down to the surface, but later analysis suggested that the crushing atmosphere damaged the spacecraft at an altitude of almost fifteen miles (twenty-five kilometers). Before its demise, it revealed an atmosphere made almost entirely of carbon dioxide, with temperatures ranging from 104°F (40°C) high up in the atmosphere to

536°F (280°C) closer to the surface. The spacecraft also experienced pressures ranging from fifteen to twenty-two atmospheres.

Mariner 5

Launch Date: June 14, 1967
Arrival at Venus: October 19, 1967
Sponsoring Country: USA
Type of Mission: Flyby of Venus
Mission Summary: *Mariner 5* flew within 2,400 miles (4,000 kilometers) of the Venusian cloud tops during its quick flyby of the planet. The spacecraft measured a surface temperature of 513°F (267°C) and a very weak magnetic field.

Cosmos 167

Launch Date: June 17, 1967
Sponsoring Country: USSR
Type of Mission: Probe
Mission Summary: The final stage of the rocket carrying the spacecraft into orbit failed and it was unable to achieve the necessary trajectory to carry it on to Venus.

Venera 5

Launch Date: January 5, 1969
Arrival at Venus: May 16, 1969
Sponsoring Country: USSR
Type of Mission: Atmospheric probe
Mission Summary: When *Venera 5* arrived at Venus, the spacecraft deployed a parachute and began its descent through the thick atmosphere. Scientists received fifty-three minutes' worth of data as the spacecraft descended twenty-two miles (thirty-six kilometers) through the Venusian clouds. Just before reaching the surface, *Venera 5* was destroyed by the tremendous atmospheric pressure.

Venera 6

Launch Date: January 10, 1969
Arrival at Venus: May 17, 1969
Sponsoring Country: USSR
Type of Mission: Atmospheric probe
Mission Summary: Identical to *Venera 5*, *Venera 6* arrived at Venus just a day behind its sister ship. Upon its arrival, the spacecraft deployed a parachute and descended through the atmosphere. Scientists on Earth received fifty-one minutes of data as the probe descended

twenty-four miles (thirty-eight kilometers). Like its sister ship, *Venera 6* was damaged by the crushing pressure before it reached the surface.

Venera 7—the first spacecraft to transmit data back from the surface of another planet

Launch Date: August 17, 1970

Arrival at Venus: December 15, 1970

Sponsoring Country: USSR

Type of Mission: Lander

Mission Summary: When *Venera 7* arrived at Venus, it deployed a parachute and began its descent to the surface. Scheduled to take one hour to descend, the probe touched down in only thirty-five minutes, possibly because of a damaged parachute. The spacecraft then transmitted a weak signal for twenty-three minutes, becoming the first spacecraft to return data from the surface of another planet. It reported surface temperatures of 890°F (475°C) and pressures ninety times greater than Earth's.

Cosmos 359

Launch Date: August 22, 1970

Sponsoring Country: USSR

Type of Mission: Lander

Mission Summary: The final stage of the rocket carrying the spacecraft into orbit failed and it was unable to achieve the necessary trajectory to carry it on to Venus.

Venera 8

Launch Date: March 27, 1972

Arrival at Venus: July 22, 1972

Sponsoring Country: USSR

Type of Mission: Lander

Mission Summary: Soon after *Venera 8* arrived at Venus, the atmospheric drag slowed the spacecraft as it entered the upper atmosphere. Then a parachute was deployed. A refrigeration unit was used to cool the spacecraft's components, protecting it from the intense heat as it descended to the surface. Once on the ground, the spacecraft transmitted data for fifty minutes, confirming a very high surface temperature and crushing atmospheric pressure.

Cosmos 482

Launch Date: March 31, 1972

Sponsoring Country: USSR

Type of Mission: Lander

Mission Summary: The final stage of the rocket carrying the spacecraft into orbit failed and it was unable to achieve the necessary trajectory to carry it on to Venus.

Mariner 10

Launch Date: November 3, 1973

Arrival at Venus: February 5, 1974

Sponsoring Country: USA

Type of Mission: Venus gravity assist for Mercury flyby

Mission Summary: *Mariner 10* used the gravity of Venus to alter its course and speed. This adjustment was necessary to allow the spacecraft to continue on toward its primary target, the planet Mercury. While flying within 2,600 miles (4,200 kilometers) of Venus, *Mariner 10* took the first ultraviolet images of the planet.

Venera 9—the first spacecraft to send back images from the surface of another planet

Launch Date: June 8, 1975

Arrival at Venus: October 22, 1975

Sponsoring Country: USSR

Type of Mission: Orbiter and lander

Mission Summary: The *Venera 9* lander separated from its orbiter on October 20, 1975. Two days later, the lander touched down and became the first spacecraft to transmit a picture from the surface of another planet. In addition, the lander sent back information on the Venusian clouds, atmospheric composition, and light levels. All the information was transmitted from the surface to the orbiter, which then relayed the signal to Earth. The lander was able to transmit data for fifty-three minutes before the orbiter moved out of range of its signal. Besides acting as a data relay, the orbiter also studied the cloud structure of the planet.

Venera 10

Launch Date: June 14, 1975

Arrival at Venus: October 25, 1975

Sponsoring Country: USSR

Type of Mission: Orbiter and lander

Mission Summary: The *Venera 10* spacecraft separated into two different sections, an orbiter and a lander, on October 23, 1975. Two days later, the lander touched down on the surface of Venus 1,370 miles (2,200 kilometers) from the *Venera 9* lander. With the orbiter acting as a relay, the *Venera 10* lander transmitted images from the surface as well as data about clouds and the surface environment. The lander transmitted sixty-five minutes' worth of data before the orbiter moved out of range of its signal.

Pioneer Venus 1

Launch Date: May 20, 1978

Arrival at Venus: December 4, 1978

Sponsoring Country: USA

Type of Mission: Orbiter

Mission Summary: *Pioneer Venus 1* carried seventeen experiments onboard, including a surface radar mapper. While in orbit around the planet, scientists used the spacecraft's radar to map nearly all of the planet's surface, resolving features as small as fifty miles (eighty kilometers). The spacecraft remained in orbit until October 8, 1992, when it used up all its fuel and burned up in the atmosphere.

Pioneer Venus 2

Launch Date: August 8, 1978

Arrival at Venus: December 9, 1978

Sponsoring Country: USA

Type of Mission: Probe

Mission Summary: *Pioneer Venus 2* consisted of four separate atmospheric probes: one large probe (five feet, or one and a half meters, in diameter), which deployed a parachute to slow its descent, and three small probes (two and a half feet, or just under one meter, across), which plunged straight through the atmosphere. The large probe was released from the spacecraft on November 16, 1978, while the spacecraft was still almost a month away from Venus. The three smaller probes were released four days later, and all the probes arrived at the planet on December 9, 1978. Each probe took atmospheric measurements as it descended through the cloud layer. One of the probes survived to transmit data for over an hour after it impacted with the surface.

Venera 11

Launch Date: September 9, 1978

Arrival at Venus: December 25, 1978

Sponsoring Country: USSR

Type of Mission: Orbiter and lander

Mission Summary: When *Venera 11* arrived at Venus, it immediately descended through the planet's dense atmosphere and soft-landed on the Venusian surface. Details about this mission are sketchy; however, the spacecraft did send back evidence of thunder and lightning as well as the presence of carbon monoxide in the lower altitudes. Data was transmitted back to Earth for ninety-five minutes before the lander rotated out of range of the orbiting relay.

Venera 12

Launch Date: September 14, 1978

Arrival at Venus: December 21, 1978

Sponsoring Country: USSR

Type of Mission: Orbiter and lander

Mission Summary: Launched three days after *Venera 11*, *Venera 12* actually made it to Venus four days before the other spacecraft, separating from its main bus on December 19, 1978, and soft-landing on the surface of the planet on December 21. *Venera 12* was designed to study the atmospheric composition and clouds of Venus. The lander transmitted 110 minutes of data before the planet rotated out of range of the orbiting relay.

Venera 13

Launch Date: October 30, 1981

Arrival at Venus: March 1, 1982

Sponsoring Country: USSR

Type of Mission: Orbiter and lander

Mission Summary: *Venera 13* soft-landed on Venus four days before its "twin," *Venera 14*. Images were sent back from the surface, and a drilling arm collected a sample that was examined by an onboard x-ray fluorescence spectrometer to determine its composition. The lander survived for 127 minutes before giving in to the extreme heat (855°F or 457°C) and the tremendous pressure (eighty-four times the pressure at sea level on Earth).

Venera 14

Launch Date: November 4, 1981

Arrival at Venus: March 5, 1982

Sponsoring Country: USSR

Type of Mission: Orbiter and lander

Mission Summary: *Venera 14* soft-landed on the surface of Venus 590 miles (950 kilometers) southwest of *Venera 13*, which had landed on the planet four days earlier. The lander sent back images of the surface and a mechanical arm collected a sample for testing. The spacecraft survived for fifty-seven minutes before succumbing to the heat and extreme pressure.

Venera 15

Launch Date: June 2, 1983

Arrival at Venus: October 10, 1983

Sponsoring Country: USSR

Type of Mission: Orbiter

Mission Summary: *Venera 15* and *Venera 16* (which arrived at Venus four days after *Venera 15*) worked together for eight months to create a radar map of Venus.

Venera 16

Launch Date: June 7, 1983

Arrival at Venus: October 10, 1983

Sponsoring Country: USSR

Type of Mission: Orbiter

Mission Summary: *Venera 16* and *Venera 15* (which arrived at Venus four days earlier) worked together for eight months to create a radar map of Venus.

Vega 1—the first balloon to explore another planet

Launch Date: December 15, 1984

Arrival at Venus: June 11, 1985

Sponsoring Country: USSR

Type of Mission: Venus probe and Halley's comet flyby

Mission Summary: *Vega 1* flew by Venus on its way to Halley's comet. As the spacecraft swung by the planet, it deployed an eight-foot (two-and-a-half-meter) probe into the atmosphere. The probe itself deployed a balloon almost immediately upon entering the atmosphere. The probe took readings of the atmosphere as it descended to the surface, while the balloon measured temperature, pressure, wind velocity, and visibility as it floated above the Venusian surface. The balloon survived forty-seven hours before it burst. During that time, it traveled over 5,590 miles (9,000 kilometers).

Vega 2

Launch Date: December 21, 1984

Arrival at Venus: June 15, 1985

Sponsoring Country: USSR

Type of Mission: Venus probe and Halley's comet flyby

Mission Summary: *Vega 2* flew by Venus on its way to Halley's comet. As the spacecraft swung by the planet, it deployed an eight-foot (two-and-a-half-meter) probe into the atmosphere. The probe itself deployed a balloon almost immediately upon entering the atmosphere. The probe took readings of the atmosphere as it descended to the surface, while the balloon measured the temperature, pressure, wind velocity, and visibility of the atmosphere. The balloon traveled around the nighttime side of Venus for two days before crossing over into daylight. Heated gases in the balloon expanded and the balloon soon burst.

Magellan—the first spacecraft to use aerobraking to change its orbit
Launch Date: May 4, 1989
Arrival at Venus: August 10, 1990
Sponsoring Country: USA
Type of Mission: Orbiter
Mission Summary: *Magellan's* goal when it first went into orbit around Venus was to provide scientists with a detailed radar map of the planet. By the end of its mission, the spacecraft had more than accomplished that goal, mapping over 98 percent of Venus at a resolution of one hundred meters or better. *Magellan* was the first spacecraft to use aerobraking to change its orbit—a process that involves using the atmospheric drag to slow a spacecraft. Controllers lost contact with the spacecraft on October 12, 1994.

Galileo
Launch Date: October 18, 1989
Arrival at Venus: February 10, 1990
Sponsoring Country: USA
Type of Mission: Venus gravity assist for Jupiter orbiter
Mission Summary: If *Galileo* was designed as a Jupiter orbiter, you may be asking yourself why then did it fly by Venus, a planet located in the opposite direction of its target! It all has to do with the fact that the spacecraft was launched from the cargo bay of the space shuttle. Its original design had the spacecraft flying directly to Jupiter, taking just a couple of years to get to the gas giant planet. However, delays forced the original launch date to slip considerably. Then, in January 1986, the space shuttle *Challenger* was destroyed in an explosion shortly after liftoff. After the disaster, NASA implemented several changes within the shuttle program, including restrictions on the types of payloads to be carried within the cargo bays of the remaining shuttles. These guidelines restricted the use of certain types of propellants, which forced engineers to use a less-powerful rocket to send the spacecraft out of Earth orbit. Now, to gain enough speed to get to Jupiter, the spacecraft had to employ a gravity-assist flyby of Venus. *Galileo* flew within 9,500 miles (16,000 kilometers) of Venus, where its speed and course were altered enough that it could begin its journey to Jupiter. *Galileo* then flew by Earth, the asteroid Gaspra, Earth again, and the asteroid Ida before arriving at Jupiter in December 1995.

Cassini
Launch Date: October 15, 1997
Arrival at Venus: April 26, 1998 and June 24, 1999
Sponsoring Country: USA
Type of Mission: Venus gravity assist for Saturn orbiter

Mission Summary: Like *Galileo* before it, *Cassini* also used the gravity of Venus to increase its speed and alter its course for a journey to the outer solar system. Even though *Cassini* was launched from a traditional rocket instead of from the space shuttle's cargo bay (like *Galileo*), the mass of the spacecraft was such that it required an extra boost to bring it up to speed. In fact, it took several boosts. *Cassini* actually flew by Venus twice. The first time, the spacecraft buzzed the planet, coming within 176 miles (284 kilometers) of its surface. It then swung around the planet again, this time coming within 370 miles (600 kilometers) of the surface. Then, after one more gravity assist, this time from Earth, the spacecraft had finally built up enough speed to begin its journey to Saturn.

MISSIONS TO MARS

While the fourth planet in the solar system is still fairly close to Earth, the elliptical orbit of Mars sometimes makes it a challenge for spacecraft trying to catch up with this small planet. And when a spacecraft does catch up with Mars, the planet's weak gravity (a result of its small mass) does little to help pull the spacecraft into orbit.

Even if missions to Mars are challenging, the Red Planet is still worth the effort. Although Mars has a very thin, poisonous atmosphere and an average temperature of –220°F (–140°C), the conditions on this planet are less harsh than on any other planet in the solar system (with the exception of Earth, of course). If humans are going to colonize a planet beyond Earth, Mars would be a good place to start.

Interest in traveling to the Red Planet has been strong for at least a century. Beginning, perhaps, with Percival Lowell and his infamous belief in the existence of canals and life on Mars, the Red Planet has captured the imagination of the masses. And even though close-up

The Mars Exploration Rover *Opportunity* took one final look at its landing site before setting off to explore other parts of Meridiani Planum. Note the rover tracks marking *Opportunity*'s trail. (Its lander is in the center of the crater.) Image credit: NASA/JPL

images of a barren, dust-covered world have dashed all hopes of finding evidence of a Martian civilization, the missions to Mars have revealed a truly unique planet with a fascinating past.

As you read through the missions, here are some interesting milestones to look for:

- the debris from this mission failure was thought to be a missile attack from the Soviet Union (*Korabl 11* or *Sputnik 22*)
- the first spacecraft to fly by Mars (*Mariner 4*)
- the first spacecraft to go into orbit around another planet (*Mariner 9*)
- the first spacecraft to successfully land on Mars (*Viking 1*)
- the mission that placed the first rover on another planet (*Mars Pathfinder*)

THE MISSIONS

Korabl 4 (Marsnik 1)

Launch Date: October 10, 1960

Sponsoring Country: USSR

Type of Mission: Flyby of Mars

Mission Summary: *Korabl 4* was the Soviet Union's first attempt at an interplanetary probe. After launch, the rocket's third stage failed to lift the probe into orbit.

Korabl 5 (Marsnik 2)

Launch Date: October 14, 1960

Sponsoring Country: USSR

Type of Mission: Flyby of Mars

Mission Summary: Just after launch, the rocket's third stage failed and the probe never obtained Earth orbit.

Korabl 11 (Sputnik 22)—the debris from the mission's launch failure was thought to be missile attack from the Soviet Union

Launch Date: October 24, 1962

Sponsoring Country: USSR

Type of Mission: Flyby of Mars

Mission Summary: The spacecraft broke apart after reaching Earth orbit. The debris reentered Earth's atmosphere and was tracked by the US Ballistic Missile Early Warning System in Alaska, which first thought it was a Soviet ICBM attack in response to the Cuban missile crisis that was occurring at the time. There were a few tense moments before controllers understood what they were seeing.

Mars 1 (Sputnik 23)

Launch Date: November 1, 1962

Sponsoring Country: USSR

Type of Mission: Flyby of Mars

Mission Summary: Controllers lost contact with *Mars 1* while the spacecraft was on its way to the Red Planet. The spacecraft was about 66 million miles (107 million kilometers) from Earth when its signal was lost.

Korabl 13 (Sputnik 24)

Launch Date: November 4, 1962

Sponsoring Country: USSR

Type of Mission: Lander

Mission Summary: *Korabl 13* broke apart in Earth orbit during a burn to transfer the probe to a Mars trajectory.

Mariner 3

Launch Date: November 5, 1964

Sponsoring Country: USA

Type of Mission: Flyby of Mars

Mission Summary: A shield that was designed to protect the spacecraft during launch failed to release once it had reached Earth orbit. With its instruments covered and the extra weight of the shield dragging it down, the spacecraft was unable to obtain the necessary trajectory to send it on to Mars.

Mariner 4—the first spacecraft to fly by Mars

Launch Date: November 28, 1964

Arrival at Mars: July 14, 1965

Sponsoring Country: USA

Type of Mission: Orbiter

Mission Summary: *Mariner 4* was the first spacecraft to fly by Mars and obtain close-up pictures of the Red Planet. The spacecraft passed within 6,117 miles (9,844 kilometers) of the Martian surface, then took four days to transmit the data back to Earth. *Mariner 4* imaged a large, ancient crater on the planet and confirmed the existence of a thin atmosphere composed of carbon dioxide. Once it has passed Mars, *Mariner 4* continued on its way, returning data until October 1965, when the orientation of its antenna made communication with Earth impossible. However, scientists were able to reestablish contact with the spacecraft in late 1967 and continued to receive data until December 20, 1967, when the mission was terminated.

Zond 2

Launch Date: November 30, 1964

Sponsoring Country: USSR

Type of Mission: Flyby of Mars and probe

Mission Summary: After a successful launch, controllers lost contact with the spacecraft while *Zond 2* was on its way to Mars. *Zond 2* had just completed a midcourse correction maneuver when the signal was lost.

Mariner 6

Launch Date: February 24, 1969

Arrival at Mars: July 31, 1969

Sponsoring Country: USA

Type of Mission: Flyby of Mars

Mission Summary: *Mariner 6* successfully flew by the planet Mars, coming within 2,131 miles (3,431 kilometers) of the Martian surface at its closest approach. The spacecraft sent back a total of seventy-five images of Mars and helped to establish the mass, radius, and shape of the planet, as well as to discover that its southern polar ice cap was composed of carbon dioxide. *Mariner 6* was identical to the *Mariner 7* spacecraft, which arrived at Mars five days after *Mariner 6*.

Mariner 7

Launch Date: March 27, 1969

Arrival at Mars: August 5, 1969

Sponsoring Country: USA

Type of Mission: Flyby of Mars

Mission Summary: *Mariner 7* successfully flew within 2,131 miles (3,430 kilometers) of the Martian surface. The spacecraft took 126 images of the planet and (combined with data from *Mariner 6*) was able to help establish the mass, radius, and shape of the planet, as well as discover that its southern polar ice cap was composed of carbon dioxide. *Mariner 6* and 7 were identical spacecraft launched thirty-one days apart. They arrived at Mars within five days of each other.

Mars 1969A

Launch Date: March 27, 1969

Sponsoring Country: USSR

Type of Mission: Orbiter

Mission Summary: The third stage of the rocket launching this mission to Mars failed, caught fire, and exploded, causing the remaining pieces to come crashing back to Earth.

Mars 1969B

Launch Date: April 2, 1969

Sponsoring Country: USSR

Type of Mission: Orbiter

Mission Summary: The first stage of the rocket launching this mission to Mars failed almost immediately after liftoff and was destroyed.

Mariner 8

Launch Date: May 8, 1971

Sponsoring Country: USA

Type of Mission: Flyby of Mars

Mission Summary: Minutes after launch, the rocket's upper stage began to tumble, causing the main engine to shut down. *Mariner 8* came crashing back to Earth, landing in the Atlantic Ocean.

Cosmos 419

Launch Date: May 10, 1971

Sponsoring Country: USSR

Type of Mission: Orbiter

Mission Summary: *Kosmos 419* reached Earth orbit, but its fourth stage rocket, which would have sent the spacecraft on its way to Mars, failed to ignite. The spacecraft reentered the atmosphere and was destroyed.

Mars 2

Launch Date: May 19, 1971

Arrival at Mars: November 27, 1971

Sponsoring Country: USSR

Type of Mission: Orbiter and lander

Mission Summary: Four and a half hours before *Mars 2* arrived at the Red Planet, it released a probe that descended to the Martian surface (at a much faster rate than intended) and crashed. However, the *Mars 2* orbiter was successfully placed in an eighteen-hour orbit. The spacecraft completed 362 orbits and was shut down on August 22, 1972. Together, *Mars 2* and *3* returned sixty images of Mars, recorded temperatures ranging from −166°F to 55°F (−110°C to 13°C), produced surface relief maps, and studied the Martian gravitational and magnetic fields.

Mars 3

Launch Date: May 28, 1971

Arrival at Mars: December 2, 1971

Sponsoring Country: USSR

Type of Mission: Orbiter and lander

Mission Summary: Identical to the *Mars 2* spacecraft, *Mars 3* delivered a descent craft to the Red Planet that survived on the surface for twenty seconds before mysteriously shutting down. The *Mars 3* orbiter went into orbit around Mars. However, because of a shortage of fuel, the spacecraft was unable to obtain its planned eighteen-hour orbit. Instead, the spacecraft ended up in an orbit of almost thirteen days. *Mars 3* completed twenty orbits around Mars before it was shut down on August 22, 1972. Together, *Mars 2* and *3* returned sixty images of Mars, recorded temperatures ranging from −166°F to 55°F (−110°C to 13°C), produced surface relief maps, and studied the Martian gravitational and magnetic fields.

Mariner 9—the first spacecraft to go into orbit around another planet

Launch Date: May 30, 1971

Arrival at Mars: November 14, 1971

Sponsoring Country: USA

Type of Mission: Orbiter

Mission Summary: *Mariner 9* was the first spacecraft to go into orbit around another planet. However, excitement for its arrival was subdued by a dark cloud—literally. A Martian dust storm had started in late September and had grown to cover the entire planet. When *Mariner 9* arrived, the only surface features visible were the summit of Olympus Mons and the three volcanoes of Tharsis Ridge. Mission scientists had to wait about a month and a half before they could begin the science portion of the mission. When the spacecraft ran out of fuel on October 27, 1972, *Mariner 9* had taken a total of 7,329 images of Mars and had studied the surface composition of the planet, the density and pressure of its atmosphere, as well as the planet's gravity and topography. The spacecraft also provided the first close-up views of Phobos and Deimos, the two moons of Mars.

Mars 4

Launch Date: July 21, 1973

Arrival at Mars: February 10, 1974

Sponsoring Country: USSR

Type of Mission: Orbiter

Mission Summary: When *Mars 4* arrived at the Red Planet, its retro-rockets failed to fire. So, instead of slowing down and going into orbit around Mars, the spacecraft flew by the planet. As it flew past Mars at a distance of 1,370 miles (2,200 kilometers), it took one set of images and collected a limited amount of data.

Mars 5

Launch Date: July 25, 1973

Arrival at Mars: February 12, 1974

Sponsoring Country: USSR

Type of Mission: Orbiter

Mission Summary: After *Mars 5* successfully went into orbit around the Red Planet, the spacecraft completed twenty-two orbits in the same number of days. Then a problem onboard the spacecraft caused communications with the ground to fail, terminating the mission. Before the spacecraft shut down, it returned sixty images of Mars.

Mars 6

Launch Date: August 5, 1973

Arrival at Mars: March 12, 1974

Sponsoring Country: USSR

Type of Mission: Flyby of Mars and lander

Mission Summary: When *Mars 6* arrived at the Red Planet, its lander separated from the main spacecraft and descended through the atmosphere, transmitting almost four minutes of data before abruptly cutting off (either when the retro-rockets fired or when it slammed into the ground). Although this was the first data of its kind (from within the Martian atmosphere), most of it was garbled and unusable. The main spacecraft acted as a transmission relay between the descent craft and Earth. It also performed an occultation experiment before it flew past the Red Planet.

Mars 7

Launch Date: August 9, 1973

Sponsoring Country: USSR

Type of Mission: Lander

Mission Summary: Although *Mars 7* successfully traveled to the Red Planet, an error onboard the spacecraft caused the lander to separate early, which resulted in its missing the planet by 800 miles (1,300 kilometers).

Viking 1—the first spacecraft to land successfully on Mars

Launch Date: August 20, 1975

Arrival at Mars: June 19, 1976

Sponsoring Country: USA

Type of Mission: Orbiter and lander

Mission Summary: When *Viking 1* arrived at Mars, it went into orbit around the Red Planet and began taking pictures of the surface in search of a safe landing site for its lander. Mission planners were hoping for a July 4 landing, but the original landing site turned out to be too rocky upon closer inspection. Another site was chosen and the first successful Mars landing took place on July 20, 1976. The *Viking 1* lander touched down in an area known as the Chryse Planitia (22.48 degrees north, 49.97 degrees west). The lander took extensive weather readings as well as conducted experiments on soil samples that it had collected with a scoop. Both the orbiter and the lander took many images of the Red Planet and collected a tremendous amount of atmospheric and surface data. The orbiter was powered down on August 17, 1980, after 1,400 orbits. The lander was shut down on November 13, 1982.

Viking 2

Launch Date: September 9, 1975

Arrival at Mars: August 7, 1976

Sponsoring Country: USA

Type of Mission: Orbiter and lander

Mission Summary: On September 3, 1976, almost a month after the spacecraft went into orbit around Mars, the *Viking 2*'s lander separated from its orbiter and descended through the Martian atmosphere. It touched down in an area known as the Utopia Planitia (47.27 degrees north, 225.99 degrees west), placing *Viking 2* on the opposite side of the planet and almost 930 miles (1,500 kilometers) closer to the north pole than *Viking 1*. One of the lander's legs came to rest on a rock, so the entire lander was tilted by about eight degrees. The lander took extensive atmospheric readings as well as conducted experiments on soil samples that it had collected with a scoop. Between *Viking 1* and *Viking 2*, more than 1,400 images were taken of the Martian surface. The *Viking 2* lander quit operating on April 11, 1980, when its batteries failed. The *Viking 2* orbiter returned almost 16,000 images before being shut down on July 25, 1978, after 706 orbits.

Phobos 1

Launch Date: July 7, 1988

Sponsoring Country: USSR

Type of Mission: Orbiter

Mission Summary: *Phobos 1* was designed to study the Sun and interplanetary space while on its way to Mars. Once in orbit around the Red Planet, it was going to study Mars and take close-up images of its moon Phobos. However, on September 2, 1988, only two months into the flight, controllers on the ground accidentally uploaded software containing a command that deactivated the spacecraft's attitude control thrusters. The spacecraft then turned its solar panels away from the Sun and was unable to recharge its batteries. As a result, the mission was lost.

Phobos 2

Launch Date: July 12, 1988

Arrival at Mars: January 29, 1989

Sponsoring Country: USSR

Type of Mission: Orbiter and 2 Phobos landers

Mission Summary: *Phobos 2* was designed to orbit Mars and land two "hoppers" on the surface of Phobos, one of the planet's two moons. The spacecraft went into orbit around Mars on January 29, 1989, and began sending back preliminary data. Then, on March 27, 1989, just before the spacecraft was to move within 160 feet (50 meters) of Phobos and deploy the two landers, the spacecraft's onboard computer malfunctioned and the mission was lost.

Mars Observer

Launch Date: September 25, 1992

Sponsoring Country: USA

Type of Mission: Orbiter

Mission Summary: *Mars Observer* was designed to study the Red Planet from orbit. On August 21, 1993, only three days from Mars, all contact with the spacecraft was suddenly lost. Engineers believe that, during pressurization of the propulsion system, unwanted mixing of propellant caused an explosion. Propellant spewing from the spacecraft spun up *Mars Observer*, causing it to lose control.

Mars Global Surveyor (MGS)

Launch Date: November 7, 1996

Arrival at Mars: September 12, 1997

Sponsoring Country: USA

Type of Mission: Orbiter

Mission Summary: In an effort to save fuel, weight, and expense, engineers were planning to use a technique called aerobraking to tweak the orbit of *Mars Global Surveyor* (*MGS*) once it reached Mars. *MGS* originally went into an elliptical forty-five-hour orbit around the planet. Then, to place the spacecraft into its desired two-hour circular orbit, *MGS* was scheduled to go through a series of aerobraking maneuvers. During these maneuvers, the spacecraft would dip into the atmosphere at specific times and the resulting atmospheric drag would slow the spacecraft and alter its course. During one of the aerobraking passes, engineers on Earth noticed that one of the spacecraft's solar panels had been overextended. Faced with the possibility of losing the entire solar panel, engineers decided to dramatically slow down the aerobraking process in order to avoid stressing the weak panel. It took over a year and a half, but engineers were able to aerobrake the spacecraft gently into its proper orbit and save the mission. The spacecraft's primary mission, to take high-resolution images of the surface and to study the planet's weather, climate, surface, and atmosphere, was completed on January 31, 2001. It is now well into its extended mission and continues to return images and data from Mars. In addition to collecting data, the spacecraft acted as a communication relay for *Spirit* and *Opportunity*, NASA's two Mars Exploration Rovers during their 2004 mission.

Mars 96

Launch Date: November 16, 1996

Sponsoring Country: Russia

Type of Mission: Orbiter, lander, and two penetrators

Mission Summary: The *Mars 96* spacecraft consisted of an orbiter, two landers, and two soil penetrators. The rocket carrying the spacecraft launched successfully, but its fourth stage ignited prematurely and sent the spacecraft crashing back to Earth.

Mars Pathfinder and Sojourner—the first rover on another planet

Launch Date: December 4, 1996

Arrival at Mars: July 4, 1977

Sponsoring Country: USA

Type of Mission: Lander and rover

Mission Summary: When *Pathfinder* arrived at Mars on July 4, 1997, it passed directly through the atmosphere, opened a parachute, fired its retro-rockets, shed its heat shield, deployed a set of airbags, and then bounced to a stop on the Martian surface—all within four minutes of arriving at the Red Planet. Once on the surface, the protective airbags deflated and retracted, allowing the pyramid-shaped spacecraft to unfold its three petals and reveal a weather station and the small *Sojourner* rover. On July 6, 1997, the six-wheeled rover rolled off a ramp and onto the Martian surface. The lander (given the name *Sagan Memorial Station*) and the rover sent back Martian weather data and many images. The mission had lasted well beyond its original thirty days when all contact with the lander was lost on September 27, 1997.

Nozomi (Planet B)

Launch Date: July 3, 1998

Arrival at Mars: December 14, 2003

Sponsoring Country: Japan

Type of Mission: Orbiter

Mission Summary: Originally scheduled to arrive at Mars in October 1999, *Nozomi* failed to gain enough speed during a gravity-assist flyby of Earth on December 21, 1998. The spacecraft also used more fuel than engineers had predicted during the maneuver. The result was a spacecraft low on fuel and traveling too slow to get to Mars. In an effort to save the mission, engineers redesigned *Nozomi's* trajectory, allowing it to take a slower but more fuel-efficient route to the Red Planet. On its new course, the spacecraft was scheduled to arrive at its target in December 2003. However, during its extended journey *Nozomi* became plagued by mechanical and software problems. Within a month of reaching Mars, engineers were forced to abandon the mission when the spacecraft's main retro-rockets failed to fire. *Nozomi* was unable to slow down enough to enter a Martian orbit. Instead, the spacecraft passed within six hundred miles (one thousand kilometers) of the Martian surface on December 14, 2003.

Mars Climate Orbiter

Launch Date: December 11, 1998

Arrival at Mars: September 23, 1999

Sponsoring Country: USA

Type of Mission: Orbiter

Mission Summary: *Mars Climate Orbiter* was lost on September 23, 1999, as it attempted to go into orbit around Mars. An investigation conducted after the loss determined that a mix-up in mathematical units had placed the spacecraft too close to Mars at the time of orbital insertion. Coming in too close to the planet meant that the spacecraft was much lower in the atmosphere than it should have been. The atmospheric drag was too great and caused the spacecraft to break apart. *Mars Climate Orbiter* was going to study the Martian climate and look for signs of water.

When calculating a spacecraft's position in space, engineers must take many forces into account, including small forces that are applied to the spacecraft by its thrusters. In the case of *Mars Climate Orbiter*, these small forces were calculated by one group and then passed along to another group for use in the spacecraft's navigation. The mix-up occurred when the first group mistakenly used English units of force (pounds) in its calculations. The mission plan called for the calculations to be done in metric units, so when the second group received the data, these engineers assumed the forces had been calculated in Newtons. The result of this unit mix-up was a spacecraft arriving sixty miles (one hundred kilometers) closer to the planet than expected, placing it much deeper into the atmosphere than originally planned. Traveling much too fast for the dense air, the spacecraft probably broke apart, with some of the pieces falling to the ground and others skipping off the atmosphere and back into space.

Mars Polar Lander

Launch Date: January 3, 1999
Arrival at Mars: December 3, 1999
Sponsoring Country: USA
Type of Mission: Lander
Mission Summary: When *Mars Polar Lander* arrived at Mars, it turned its antenna away from Earth to prepare for its entry into the Martian atmosphere. This was the last time controllers heard from the spacecraft. A review board determined the most likely cause for the loss of the mission was a faulty software system that may have triggered the retro-rockets to turn off early, causing the lander to crash instead of gently touch down on the surface. The spacecraft was going to land on the layered terrain of the southern polar ice cap, less than six hundred miles (one thousand kilometers) from the south pole. It had a scoop to collect soil samples as well as a microphone to record the sounds of Mars.

Deep Space 2 (Amundsen and Scott)

Launch Date: January 3, 1999
Arrival at Mars: December 3, 1999
Sponsoring Country: USA
Type of Mission: Penetrators (designed to crash-land and bore into the surface of its target)
Mission Summary: Carried piggyback onboard the *Mars Polar Lander*, the two microprobes were designed to separate from the *Mars Polar Lander*'s cruise stage just before the spacecraft entered the Martian atmosphere. The probes were to fall to the surface, hitting the ground at speeds of 525–660 feet/second (160–200 meters/second). Their heat shields would shatter on impact and the probes would penetrate the ground by as much as a meter, depending on the composition of the soil. Data from the probes were to be sent to the orbiting *Mars Global Surveyor,* which would then relay the data back to Earth. However, no signal was ever received.

Mars Odyssey

Launch Date: April 7, 2001

Arrival at Mars: October 24, 2001

Sponsoring Country: USA

Type of Mission: Orbiter

Mission Summary: *Odyssey* arrived at Mars and, after tweaking its orbit, began its science mission in January 2002. It is capturing images of the Martian surface at resolutions between those of *Viking* and *Mars Global Surveyor*, and is making both daytime and nighttime observations of the surface in thermal infrared wavelengths at resolutions higher than ever seen before. *Odyssey's* primary mission ended in August 2004, but with all its systems still working perfectly, the spacecraft's mission has been extended.

Mars Express

Launch Date: June 2, 2003

Arrival at Mars: December 25, 2003

Sponsoring Countries: Europe

Type of Mission: Orbiter

Mission Summary: *Mars Express,* the European Space Agency's first mission to another planet, arrived at Mars on Christmas Day, 2003. Taking a couple of weeks to adjust its orbit, the spacecraft began studying various aspects of the Martian atmosphere and terrain in early January 2004. *Mars Express* is working to complete high resolution photo-geology maps, as well as maps of the mineralogy of the planet and its subsurface structure.

Beagle 2

Launch Date: June 2, 2003

Arrival at Mars: December 25, 2003

Sponsoring Countries: Europe

Type of Mission: Lander

Mission Summary: Carried piggyback onboard *Mars Express*, *Beagle 2* detached itself from the other spacecraft as the two approached Mars. It was then scheduled to bounce to a landing on the Martian surface, but no signal was ever received from the lander. *Beagle 2* carried an array of scientific instruments to study the Martian soil, atmosphere, and more.

Spirit (Mars Exploration Rover A)

Launch Date: June 10, 2003

Arrival at Mars: January 4, 2004

Sponsoring Country: USA

Type of Mission: Rover

Mission Summary: *Spirit*, the first of two Mars Exploration Rovers, successfully bounced to a landing on the Red Planet on January 4, 2004, and began sending back amazing images within hours of its arrival. The rover, after a brief computer memory problem, successfully roamed over the rocky terrain of Gusev Crater, acting as a robotic field geologist. *Spirit* broke all previous distance records set by the *Pathfinder* rover *Sojourner,* as it traveled over 2.8 miles (3.5 kilometers) to reach a series of hills named after the space shuttle *Columbia*'s last crew. With the rover in excellent health, its original ninety-day mission was extended once. Then, after *Spirit* weathered the worst of the Martian winter and engineers worked around a faulty front wheel, its mission was extended again.

Opportunity (Mars Exploration Rover B)

Launch Date: July 8, 2003

Arrival at Mars: January 25, 2004

Sponsoring Country: USA

Type of Mission: Rover

Mission Summary: *Opportunity*, the second of the two Mars Exploration Rovers, successfully bounced to a landing on Mars on January 25, 2004, coming to rest in a small crater on the otherwise flat Meridiani Planum. *Opportunity* carefully investigated its new home, which included a small outcropping of rocks unlike anything ever seen before on Mars. Within this outcropping, the rover found strong physical and chemical evidence for the existence of an ancient body of salt water at this location. *Opportunity* then left the small crater and began exploring other areas of Meridiani Planum. The rover completed its ninety-day primary mission and was granted an extension that allowed it to continue its journey of discovery. Then, even with a malfunctioning heater on the rover's arm, *Opportunity* weathered the worst of the Martian winter and its mission was extended again.

MISSIONS TO JUPITER

Jupiter, the fifth planet from the Sun, is also the largest and most massive planet in the solar system. In fact, the mass of Jupiter is greater than twice the mass of all the other planets combined. This incredible mass translates into extremely strong gravity—which is something that engineers and scientists must take into account when trying to send a spacecraft to this planet. And they have done so with tremendous success. Every mission launched toward Jupiter has made it to the planet relatively unscathed. (*Galileo* had some antenna problems, but those occurred well before the spacecraft got to Jupiter, and the spacecraft was still able to return some great data.)

These missions have revealed much about the largest and closest gas giant planet to Earth. As you read through the missions, here are some interesting milestones to look for:

- the first spacecraft to pass through the asteroid belt (*Pioneer 10*)
- the first spacecraft to visit a gas giant planet (*Pioneer 10*)
- the first spacecraft to go into orbit around Jupiter (*Galileo*)
- the mission that used Jupiter as a gravity assist to get to the Sun (*Ulysses*)

Jupiter as seen by the *Cassini* spacecraft. Image credit: NASA/JPL/University of Arizona

THE MISSIONS

Pioneer 10—the first spacecraft to cross the asteroid belt and fly by Jupiter

Launch Date: March 2, 1972

Arrival at Jupiter: December 3, 1973

Sponsoring Country: USA

Type of Mission: Flyby of Jupiter

Mission Summary: *Pioneer 10* was the first spacecraft to fly by Jupiter, passing within 124,000 miles (200,000 kilometers) of the Jovian cloud tops. Scientists were surprised at the tremendous radiation levels experienced by the spacecraft as it flew by the gas giant planet. *Pioneer 10* studied Jupiter's magnetic field and atmosphere while it took the first close-up pictures of the planet. Once past Jupiter, the spacecraft headed out of the solar system. Because of budget cuts, routine communication with *Pioneer 10* ended on March 31, 1997. However, the spacecraft was still functioning and controllers occasionally checked in with it as it headed toward Aldebaran, the red giant star in the constellation of Taurus. (At its current speed, it will take about two million years to get to Aldebaran.) The last time controllers were able to contact the distant spacecraft was on January 23, 2003.

Pioneer 11

Launch Date: April 5, 1973

Arrival at Jupiter: December 2, 1974

Sponsoring Country: USA

Type of Mission: Flyby of Jupiter and Saturn

Mission Summary: *Pioneer 11* was the second spacecraft to explore the outer solar system (the first being *Pioneer 10*). The spacecraft flew by Jupiter, coming within 21,100 miles (34,000 kilometers) of the Jovian cloud tops. *Pioneer 11* studied the planet's atmosphere and magnetic field, as well as took pictures of Jupiter and some of its moons. Once past Jupiter, the spacecraft continued on toward Saturn and then the outer reaches of the solar system. Its instruments were finally shut down on September 9, 1995, when its nuclear power supply could no longer provide enough energy to power its equipment.

Voyager 2

Launch Date: August 20, 1977

Arrival at Jupiter: July 9, 1979

Sponsoring Country: USA

Type of Mission: Flyby of the outer planets

Mission Summary: Even though *Voyager 2* was launched sixteen days before *Voyager 1*, it took its time getting to Jupiter and arrived four months after *Voyager 1*. Ten months into its flight and well before the spacecraft reached Jupiter, *Voyager 2*'s primary radio receiver failed. The backup receiver kicked in, but it proved to be somewhat unreliable. Controllers tried to revive the primary receiver, without any luck, so they were forced to continue with the backup. Despite its irregularities, the backup receiver worked admirably during the Jupiter flyby, and continues to work to this day. *Voyager 2* took almost eighteen thousand images of Jupiter and its moons. Between *Voyager 1* and *2*, three new moons were discovered as well as a thin, dark ring around Jupiter. *Voyager*'s images of Jupiter's moon Io revealed active volcanoes—the first ever discovered on another body besides Earth. *Voyager 2* continued on to Saturn, Uranus, and Neptune, and it is now heading out trying to detect the boundary between our solar system and interstellar space.

Voyager 1

Launch Date: September 5, 1977

Arrival at Jupiter: March 5, 1979

Sponsoring Country: USA

Type of Mission: Flyby of the outer planets

Mission Summary: Launched sixteen days after *Voyager 2*, *Voyager 1* was on the fast track to Jupiter and actually arrived at the gas giant planet four months ahead of the other spacecraft. When *Voyager 1* flew by Jupiter, it took more than eighteen thousand images of Jupiter and its moons (about the same number as *Voyager 2*). *Voyager 1* then continued on to Saturn. During the two *Voyager* flybys of Jupiter, three new moons were discovered as well as a thin, dark ring around the planet. The *Voyager* images of Jupiter's moon Io revealed active volcanoes—the first ever discovered on another body besides Earth.

Galileo—the first spacecraft to go into orbit around Jupiter

Launch Date: October 18, 1989

Arrival at Jupiter: December 8, 1995

Sponsoring Country: USA

Type of Mission: Orbiter

Mission Summary: *Galileo* went into orbit around Jupiter on December 8, 1995. Even though *Galileo*'s main antenna had failed to deploy entirely, the spacecraft was able to send back incredible images of Jupiter and the larger moons during its two-and-a-half-year primary mission. *Galileo*'s mission was then extended twice, with the spacecraft surviving three times longer than originally planned. The spacecraft continued in orbit around Jupiter until September 21, 2003, when it was programmed to plunge into the gas giant.

Galileo Probe (onboard *Galileo*)

Launch Date: October 18, 1989

Arrival at Jupiter: December 7, 1995

Sponsoring Country: USA

Type of Mission: Probe

Mission Summary: The *Galileo Probe* was released from the *Galileo* spacecraft six months before it reached the gas giant planet. One day before *Galileo* went into orbit around Jupiter, the *Galileo Probe* plunged into the planet's thick clouds. It was protected by a heat shield, and its descent was slowed by a parachute. The probe sent back information about the planet's temperature, wind speeds, and pressure as it descended through the clouds, using *Galileo* as a data relay. It finally succumbed to incredible pressures (twenty-four times Earth's pressure at sea level) one hour after it began its descent.

Ulysses

Launch Date: October 6, 1990

Arrival at Jupiter: February 8, 1992

Sponsoring Countries: USA and Europe

Type of Mission: Jupiter gravity assist for solar orbiter

Mission Summary: The primary mission of *Ulysses* was to study the north and south poles of the Sun. However, getting to those solar poles required the spacecraft to perform some interplanetary gymnastics. The spacecraft first went to Jupiter, where the strong Jovian gravity helped redirect the spacecraft, placing it on its proper course. As *Ulysses* flew by the planet, instruments onboard the spacecraft studied Jupiter's strong magnetic field and radiation levels.

Cassini

Launch Date: October 15, 1997

Arrival at Jupiter: December 30, 2000

Sponsoring Country: USA

Type of Mission: Saturn orbiter

Mission Summary: Designed as a Saturn orbiter, *Cassini*'s path took it past Venus, Earth, and Jupiter in a roundabout journey to the ringed planet Saturn. Engineers used the Jupiter encounter to test the spacecraft's instruments and operations. During the flyby, *Cassini* captured some amazing images of the gas giant and its large moons.

MISSIONS TO SATURN

Saturn is the fifth planet from the Sun. At an average distance of 887.4 million miles (1.4 billion kilometers), the planet is almost twice as far from the Sun as Jupiter is. Light traveling this distance takes over an hour to reach the planet, and when it does, it is only about a quarter of the strength it was at Jupiter. Traveling this distance is one of the many difficulties faced by spacecraft trying to reach Saturn. Years of interplanetary travel can be hard on a spacecraft. Gears can jam, transmitters can fail, stabilizing gyros can go bad. About the only thing engineers and scientists on Earth can do to avoid problems during flight is to build the spacecraft with the most durable materials available, make backups of crucial systems, and test everything as much as possible before launch.

Once a spacecraft gets to Saturn, it faces another challenge, the planet's beautiful ring system. Jupiter, Uranus, and Neptune also have rings, but none of them are as vast and complex as Saturn's. Particles within Saturn's rings range in size from a grain of sand to boulders as large as a house. As these particles travel around the planet, their orbital velocity turns even the smallest grain of dust into a potential spacecraft killer. So, to avoid any danger of collision with ring particles, engineers and scientists program the spacecraft to stay clear of the area immediately surrounding Saturn, limiting how

The *Cassini* spacecraft took this image of Saturn and its rings as it approached the planet. Image credit: NASA/JPL/Space Science Institute

289

close a spacecraft can get to the planet—although the *Cassini* spacecraft actually flew through a gap between the two most distant rings. Even if the rings prevent us from getting too close, spacecraft visiting Saturn have gotten some great views.

As you read through the missions, here are some interesting milestones to look for:

- the first spacecraft to fly by Saturn (*Pioneer 11*)
- the first spacecraft to go into orbit around Saturn (*Cassini*)

THE MISSIONS

Pioneer 11—the first spacecraft to fly by Saturn

Launch Date: April 5, 1973

Arrival at Saturn: September 1, 1979

Sponsoring Country: USA

Type of Mission: Flyby of Jupiter and Saturn

Mission Summary: *Pioneer 11* was the first spacecraft to visit the ringed world of Saturn, flying within 13,000 miles (21,000 kilometers) of the planet's cloud tops. During the flyby, *Pioneer 11* studied Saturn's atmosphere and magnetic field and took pictures of the planet and its moons. Once past Saturn, *Pioneer 11* continued on, traveling out of the solar system until its nuclear power supply was unable to provide enough energy to run its instruments. The spacecraft was shut down on September 30, 1995—over twenty-two years after its journey began.

Voyager 2

Launch Date: August 20, 1977

Arrival at Saturn: August 26, 1981

Sponsoring Country: USA

Type of Mission: Flyby of the outer planets

Mission Summary: (*Voyager 2* was launched sixteen days before *Voyager 1*. However, *Voyager 1* was on a faster route and made it to the ringed world first.) *Voyager 2* flew within 25,400 miles (41,000 kilometers) of Saturn's cloud tops and provided scientists with almost sixteen thousand images of the planet and its moons and rings. While at Saturn, the two *Voyager* spacecraft discovered three new moons as well as the intricate structure and spoke-like features of the ring system, and collected information about the planet's atmosphere and magnetic field. *Voyager 2* used Saturn's strong gravity to change its course and increase its speed so it was able to fly on to Uranus and Neptune.

Voyager 1

Launch Date: September 5, 1977

Arrival at Saturn: November 12, 1980

Sponsoring Country: USA

Type of Mission: Flyby of the outer planets

Mission Summary: *Voyager 1* came within 40,000 miles (64,200 kilometers) of Saturn's cloud tops during its flyby of the planet. The spacecraft took almost sixteen thousand images of Saturn, its moons, and its ring system. When *Voyager 1* flew by the planet, controllers on the ground directed the spacecraft to use Saturn's strong gravity to alter its course so that it flew up and out of the plane in which the planet orbits. As the spacecraft moved up and away, its cameras looked down on the ringed world, providing scientists with new views of Saturn and its rings. *Voyager 1* is currently on an interstellar mission and is the most distant human-made object ever launched, taking that title from *Pioneer 10* on February 17, 1998. The spacecraft is now traveling at a speed of 335 million miles (365 million kilometers) per year. In July 2005, *Voyager 1* will be 8.8 billion miles (14.3 billion kilometers) from the Sun.

Cassini and the Huygens Probe

Launch Date: October 15, 1997

Arrival at Saturn: July 1, 2004

Sponsoring Countries: USA and Europe

Type of Mission: Orbiter and Moon probe

Mission Summary: The *Cassini Orbiter* was designed to tour Saturn and its icy moons. The spacecraft's voyage to the ringed world was a long and complicated one that included two Venus flybys, an Earth flyby, and a Jupiter flyby before it even came close to its target. Then, on June 11, 2004, when the spacecraft was less than a month from Saturn, it flew within 1,240 miles (2,000 kilometers) of Phoebe, one of Saturn's most distant moons, providing scientists with their first ever images of this tiny rocky world. When *Cassini* arrived at Saturn on July 1, 2004, it first passed up through a gap between the planet's F and G rings, then came within 12,420 miles (20,000 kilometers) of Saturn's cloud tops. Scientists were able to collect incredibly detailed images of the planet and its rings before the spacecraft flew back between the F and G rings, with its course now sufficiently altered by Saturn's gravity to enter into a stable orbit around the ringed world. During this time, the spacecraft also took measurements of the planet's magnetic field and gave scientists an opportunity to study lightning within Saturn's upper atmosphere. The spacecraft also measured impacts as it passed through the gap in the rings in an attempt to determine the quantity, density, and size of ring material. After six months in orbit around Saturn, the *Huygens* probe (designed and constructed by the European Space Agency) separated from the *Cassini* orbiter and began its entry into the atmosphere of Titan, Saturn's largest moon, offering scientists their first look at the haze-covered moon. *Cassini* is expected to make at least thirty orbits of the planet, each optimized for a different set of observations.

MISSIONS TO URANUS

How do you get to a distant world such as Uranus? You take advantage of once-in-a-lifetime opportunities. In the late 1970s and the 1980s, a rare planetary alignment provided aerospace engineers with such an opportunity. The gas giant planets were aligning themselves in almost a straight line—something that happens only once every 177 years or so. If engineers could launch a spacecraft at the proper time, they would be able to fly by each of the four planets with only minor course corrections. Four planets for the price of one spacecraft—a bargain any way you look at it. Thus, the Voyager project was born. Two identical spacecraft, *Voyager 1* and *Voyager 2*, were built and launched within a month of each other.

The two spacecraft flew by Jupiter in 1979 and Saturn in 1980–81. In order to get as many different views as possible of Saturn's impressive ring system, engineers directed *Voyager 1* to

alter its course, sending it up and out of our solar system. *Voyager 2*, however, continued on its original path. Four and a half years later, after a total of eight and a half years in space, *Voyager 2* flew by the planet Uranus. Since there are no current plans to send another spacecraft to this distant world, the *Voyager 2* results are the best we will see of this planet for at least a generation. Below is a brief mission summary of the spacecraft's Uranus encounter.

Uranus as seen by *Voyager 2*.
Image credit: NASA/JPL

293

THE MISSION

Voyager 2

Launch Date: August 20, 1977

Arrival at Uranus: January 24, 1986

Sponsoring Country: USA

Type of Mission: Flyby of the outer planets

Mission Summary: *Voyager 2* flew by Uranus on January 24, 1986, coming within 50,600 miles (81,500 kilometers) of the planet's cloud tops. The spacecraft took almost eight thousand images of the planet, its moons, and its dark ring system. The planet itself didn't offer much to look at since it was covered with a thick layer of greenish blue methane haze. However, scientists discovered that the planet has an unusually tilted axis of rotation—ninety-eight degrees compared to Earth's twenty-three-and-a-half-degree tilt. The five largest moons of Uranus also proved to be quite interesting. The moon Miranda proved to be a jumble of ice and rock with a large chevron-shaped feature whose cliffs jut up almost twelve miles (twenty kilometers) above the moon's surface. *Voyager 2* discovered an additional ten moons while flying by the planet. (Recent Earth-based discoveries have brought the Uranian moon count to twenty-seven.) The spacecraft also took images of Uranus's very dark, narrow ring system—eleven rings in all. After flying by Uranus, the spacecraft made its way to Neptune.

MISSIONS TO NEPTUNE

Because of its tremendous distance from the Sun, Neptune has been visited by only one spacecraft during the entire history of the space age. And, unfortunately, there are no new Neptune missions waiting in the wings. At an average distance of 2.98 billion miles (4.5 billion kilometers) from the Sun, it takes sunlight over four hours to reach this distant world. *Voyager 2*, the one earthly visitor to the planet, took twelve years to get there even after using gravity assists from Jupiter, Saturn, and Uranus to increase its velocity.

That long of a journey in the harsh environment of outer space takes a toll on the spacecraft as well as on human emotions. In this day of instant gratification, it's hard to imagine launching a spacecraft that will take that long to get to its target. So *Voyager 2* may well be the only visitor to this distant world in the foreseeable future. In any case, we were extremely lucky that this little spacecraft performed beyond all expectations and revealed an amazing royal blue planet, alive with dark storms and icy white clouds. Below is a brief mission summary of *Voyager 2*'s Neptune encounter.

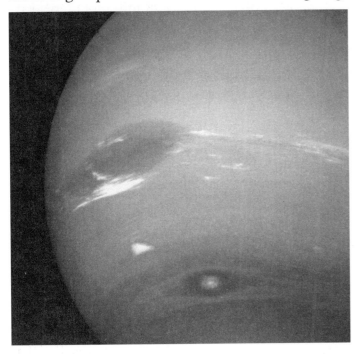

Neptune as seen by *Voyager 2*.
Image credit: NASA/JPL

THE MISSION

Voyager 2

Launch Date: August 20, 1977

Arrival at Neptune: August 25, 1989

Sponsoring Country: USA

Type of Mission: Flyby of the outer planets

Mission Summary: *Voyager 2* flew by Neptune on August 25, 1989. Scientists programmed the spacecraft to fly within 3,105 miles (5,000 kilometers) of the planet's cloud tops, closer than it had come to any previous planet. Even at such a great distance from the Sun, with the four-hour time lag in communications and the low lighting conditions, the spacecraft returned ten thousand images of Neptune, its moons, and its ring system. *Voyager 2* discovered interesting features on the planet, including the "Great Dark Spot" (an Earth-sized hurricane) and "Scooter" (a cloud formation that rotated rapidly around the planet). The spacecraft also recorded some of the fastest winds in the solar system on Neptune. Immediately surrounding the planet, the spacecraft discovered that one of Neptune's two main rings was not uniform in its distribution of material. In other words, the ring contains clumps. Beyond the rings, *Voyager 2* discovered six new moons, bringing Neptune's moon count up to eight. (In recent years, five more moons have been discovered by Earth-based observers, bringing Neptune's moon count to thirteen.)

Voyager 2 is currently heading out of our solar system on an extended interstellar mission. The spacecraft is traveling at a speed of 307 million miles (498 million kilometers) per year. While most of its instruments have been turned off, scientists are still using the spacecraft in an effort to detect an area of space known as the heliopause, where the Sun's influence on space ends and interstellar space begins. By July 2005, the spacecraft will be approximately 7.1 billion miles (11.5 billion kilometers) from the Sun.

MISSIONS TO PLUTO

As of 2005, no spacecraft has traveled to the ninth planet in our solar system. However, there are tentative plans to launch a *New Horizons Pluto–Kuiper Belt* mission in January 2006 that would arrive at Pluto in July 2015.

Many people are anxious for this mission to launch. Several Pluto missions have been proposed throughout the years, from orbiters to flybys. But all were canceled or delayed indefinitely because of high costs, technical issues, and/or lack of support stemming from the fact that the spacecraft will take a long time to get to its target.

If the *New Horizons* mission doesn't launch as predicted, we may miss a once-in-a-lifetime chance to study Pluto's atmosphere. Having reached perihelion (its closest point to the Sun) in 1989, Pluto is now heading back to the far reaches of the solar system. With each day, it is moving farther and farther from the Sun, and therefore getting colder and colder. Soon, it will be too cold to support its thin atmosphere, and the gases will freeze, coating the planet with a layer of ice. The atmosphere will remain an icy layer on the surface until Pluto's orbit brings it close enough for the Sun's heat to melt the ice again. That won't happen for another two hundred years.

In 2015, if everything goes according to plan, Pluto will lose the distinction of being the only planet in our solar system yet to be explored by a spacecraft. It may be worth calling your local congressional representative to show your support.

AFTERWORD

Space exploration is about more than seeing stunning images of distant worlds. It's about getting to know the neighborhood in which we live. The more we know about the planets in our solar system, the better we understand our own world. The more we know about our own planet, the better chance we have of overcoming obstacles such as global warming and ozone depletion.

However, space exploration is not an easy thing to accomplish. (If it were, there would already be fast-food restaurants on Mars.) For every success, there have been failures. In each case, engineers and scientists have learned from their mistakes and moved on. But how long will this be allowed to continue?

It seems as if today's society is becoming less tolerant of failures. News reports spend hours focusing on the financial cost of a lost rocket or a failed parachute, questioning whether we should be spending all of that money on other "less risky" programs. What people seem to be forgetting is that risk is a part of being human, whether it involves walking across the street or sending a spacecraft to another world.

Many of our ancestors believed that the risk was acceptable when they decided to cross the oceans to come to the New World—despite the enormous numbers of ships that were lost along the way. Without sacrifice, without taking risks, our nation would cease to grow. We would become stagnant, with nothing to look forward to but the same old grind. In addition to all of the knowledge we gain from it, space exploration—be it robotic or manned—gives us something to look forward to. It allows us to leave the day-to-day routine and gives us a glimpse of what lies beyond.

Unfortunately for some, the quest for knowledge isn't enough to justify the cost of a space program. To these people, I would like to point out a down-to-Earth reason for exploring space: technology. Much of our "modern" world owes its existence to the space program.

Check out NASA's Spinoff Web site at http://www.sti.nasa.gov/tto. You will be amazed to discover just how much technology developed specifically for the space program touches our everyday lives.

While these everyday technological advances provide a financial and practical reason for the space program to exist, the bottom line is that it is human nature to explore. Exploration is about learning. The more we explore, the more we learn.

May we continue to explore. It is well worth the cost and the risk.

NOTES

ABOUT THE SOLAR SYSTEM

1. Donald Savage and Jane Platt, "Most Distant Object in Solar System Discovered," NASA press release 04-091, March 15, 2004.

2. Voyager Web Site, "Voyager Interstellar Mission," NASA, Jet Propulsion Laboratory, Caltech, http://voyager.jpl.nasa.gov/mission/interstellar.html (accessed January 14, 2003).

3. Voyager Web Site, "Voyager at 90 AUs . . . and Beyond," NASA, Jet Propulsion Laboratory, Caltech, http://voyager.jpl.nasa.gov/index.html (accessed July 23, 2004).

4. Voyager Web Site, "Interstellar Science," NASA, Jet Propulsion Laboratory, Caltech, http://voyager.jpl.nasa.gov/science/index.html (accessed January 14, 2003).

5. Kimm Groshong, "Caltech Scientist Helps Track Down Unique Object," *Pasadena Star News*, March 16, 2004.

6. John Davies, *Beyond Pluto: Exploring the Outer Limits of the Solar System* (Cambridge: Cambridge University Press, 2001), p. 11.

7. Ibid., p. 14.

8. Roger A. Freedman and William J. Kaufmann III, *Universe*, 6th ed. (New York: W. H. Freeman and Company, 2002), p. 76.

9. David Halliday and Robert Resnick, *Fundamentals of Physics*, 2nd ed. (New York: John Wiley & Sons, 1981), p. 255.

10. Minor Planet Center, "Unusual Minor Planets," Smithsonian Astrophysical Observatory, International Astronomical Union, http://cfa-www.harvard.edu/iau/lists/Unusual.html (accessed September 17, 2004).

11. Davies, *Beyond Pluto*, p. 68.

12. Chad Trujillo, "Frequently Asked Questions about 2004 DW," http://www.gps.caltech.edu/~chad/2004dw/ (accessed September 17, 2004).

13. Amir Alexander, "Giant Planetoid Detected at Edge of Solar System," Planetary Society, http://planetary.org/news/2004/2004dw/html (accessed February 20, 2004).

14. Davies, *Beyond Pluto*, p. 84.

15. Groshong, "Caltech Scientist Helps Track Down Unique Object."

16. Savage and Platt, "Most Distant Object."

17. Ibid.

18. Jay M. Pasachoff, *Astronomy: From the Earth to the Universe* (Belmont, CA: Thomson Learning, 2002), p. 335.

19. Davies, *Beyond Pluto*, p. 19.

20. Pasachoff, *Astronomy*, p. 335.

21. Ibid., p. 336.

22. Savage and Platt, "Most Distant Object."

ABOUT THE SUN

1. Roger A. Freedman and William J. Kaufmann III, *Universe*, 6th ed. (New York: W. H. Freeman and Company, 2002), p. 181.

2. Ibid.

3. Robert Burnham Jr., *Burnham's Celestial Handbook: An Observer's Guide to the Universe beyond the Solar System*, 3 vols. (New York: Dover Publishing, 1978), 1:549.

4. Jeffrey Bennett et al., *The Solar System: The Cosmic Perspective*, 3rd ed. (San Francisco: Addison Wesley, 2004), p. A-15.

5. Freedman and Kaufmann, *Universe*, p. 397.

6. Bennett et al., *The Solar System*, p. 509.

7. Ibid.

8. Voyager Web Site, "Voyager at 90 AUs . . . and Beyond," NASA, Jet Propulsion Laboratory, Caltech, http://voyager.jpl.nasa.gov/index.html (accessed July 23, 2004).

9. Freedman and Kaufmann, *Universe*, pp. 406–407.

10. Helen Wright, *Explorer of the Universe: A Biography of George Ellery Hale* (Woodbury, NY: American Institute of Physics, 1994), p. 212.

11. Jay M. Pasachoff, *Astronomy: From the Earth to the Universe* (Belmont, CA: Thomson Learning, 2002), p. 416.

12. Ibid.

13. David H. Hathaway, "Coronal Mass Ejections," Solar Physics, Science Directorate, NASA Marshall Space Flight Center, http://science.nasa.gov/ssl/pad/solar/cmes.htm (accessed January 6, 2003).

14. National Oceanic and Atmospheric Administration (NOAA), "Sun Erupts with Intense Activity," http://www .noaanews.noaa.gov/stories2003/s2104.htm (accessed October 22, 2003).

15. David Whitehouse, "Solar Storm Surge Not Over Yet," BBC News, http://news.bbc.co.uk/1/hi/sci/tech/3230807.stm (accessed October 31, 2003).

16. Philip C. Knocke (Mars Exploration Rover Entry and Descent Mission Engineer), in discussion with the author, November 2003.

17. Robert Roy Britt, "Fastest Space Storm on Record Reaches Edge of Solar System," Space.com, http://www.space .com/scienceastronomy/solar_storms_sun_040708.html (accessed July 8, 2004).

18. Freedman and Kaufmann, *Universe*, p. 405.

19. Bennett et al., *The Solar System*, p. 497.

20. Freedman and Kaufmann, *Universe*, p. 482.

21. Pasachoff, *Astronomy*, p. 523.

22. Freedman and Kaufmann, *Universe*, p. 483.

23. Pasachoff, *Astronomy*, p. 525.

24. Freedman and Kaufmann, *Universe*, p. 485.

25. Ibid.

26. Pasachoff, *Astronomy*, p. 525.

27. Freedman and Kaufmann, *Universe*, p. 502.

28. Pasachoff, *Astronomy*, p. 526.

29. Freedman and Kaufmann, *Universe*, p. 505.

30. Pasachoff, *Astronomy*, p. 533.

31. Fred Espenak, "Solar Eclipse Page," NASA Goddard Space Flight Center, http://sunearth.gsfc.nasa.gov/eclipse/solar .html (accessed July 29, 2004).

ABOUT THE PLANETS

1. David Bellingham, *An Introduction to Greek Mythology* (Secaucus, NJ: Chartwell Books, 1989), p. 114.

2. Jay M. Pasachoff, *Astronomy: From the Earth to the Universe* (Belmont, CA: Thomson Learning, 2002), p. A-4.

3. National Space Science Data Center (NSSDC), "The Apollo Program (1963–1972)," http://nssdc.gsfc.nasa.gov/ planetary/lunar/apollo.html (accessed September 10, 2004).

4. NSSDC Master Catalog, "Mariner 10," http://nssdc .gsfc.nasa.gov/nmc/tmp/1973-085A.html (accessed September 9, 2004).

5. Mars Exploration Rover Mission, "Summary," NASA, Jet Propulsion Laboratory, Caltech, http://marsrovers.jpl.nasa .gov/overview/ (accessed July 23, 2004).

6. Jet Propulsion Laboratory's Voyager Mission page, "Jupiter," NASA, Jet Propulsion Laboratory, Caltech, http:// voyager.jpl.nasa.gov/science/jupiter.html (accessed January 14, 2003); "Saturn," NASA, Jet Propulsion Laboratory, Caltech, http://voyager.jpl.nasa.gov/science/saturn.html (accessed January 14, 2003); "Uranus," NASA, Jet Propulsion Laboratory, Caltech, http://voyager.jpl.nasa.gov/ science/uranus.html (accessed March 5, 2004); "Neptune," NASA, Jet Propulsion Laboratory, Caltech, http://voyager.jpl.nasa.gov/ science/neptune.html (accessed January 14, 2003).

7. Information derived from several sources: Roger A. Freedman and William J. Kaufmann III, *Universe*, 6th ed. (New York: W. H. Freeman and Company, 2002), pp. 181, 227, 242, 262, 285, 324, 344–45, 357; Jeffrey Bennett et al., *The Solar System: The Cosmic Perspective*, 3rd ed. (San Francisco: Addison Wesley, 2004), p. A-15; Pasachoff, *Astronomy*, pp. 305, A-4.

8. Bennett et al., *The Solar System*, p. A-16.

9. Henry L. Giclas, *The 13-Inch Pluto Discovery Telescope* (Flagstaff, AZ: Lowell Observatory, 1997), p. 9.

10. Julio Magalhaes, "Galileo Probe Mission Science Summary," NASA, Ames Research Center, Galileo Probe Mission Updates, http://spaceprojects.arc.nasa.gov/Space_Projects/galileo _probe/htmls/Science_summary.html (accessed May 10, 1996).

11. Freedman and Kaufmann, *Universe*, p. 357.

12. Pasachoff, *Astronomy*, p. 193.

13. Freedman and Kaufmann, *Universe*, p. 227.

14. George S. Brady and Henry R. Clauser, *Materials Handbook*, 11th ed. (New York: McGraw-Hill, 1979), p. 422.

15. Bill Yenne, *Atlas of the Solar System* (New York: Bison Books, 1987), p. 83.

16. *World Book Encyclopedia*, 1976 ed., s.v. "Grand Canyon."

17. Ibid., s.v. "Lake Tanganyika."

18. Pasachoff, *Astronomy*, p. 210.

19. Ronald Greeley and Raymond Batson, *NASA Atlas of the Solar System* (Cambridge: Cambridge University Press, 1997), p. 122.

20. Scott S. Shepard, "The Jupiter Satellite Page," University of Hawaii, Institute for Astronomy, http://www.ifa.hawaii .edu/~sheppard/satellites/ (accessed September 18, 2004).

21. Greeley and Batson, *NASA Atlas of the Solar System*, p. 172.

22. NSSDC, "Jovian Satellite Fact Sheet," http://nssdc .gsfc.nasa.gov/planetary/factsheet/joviansatfact.html (accessed September 14, 2004).

23. Freedman and Kaufmann, *Universe*, p. 304.

24. Guy Webster, "Solar System's Largest Moon Likely Has a Hidden Ocean," Jet Propulsion Laboratory, NASA, Caltech press release, December 16, 2000.

25. Douglas Isbell and Jane Platt, "Jupiter's Moon Callisto May Hide Salty Ocean," Jet Propulsion Laboratory, NASA, Caltech press release, October 21, 1998; Jane Platt, "Galileo Findings Boost Idea of Other-Worldly Ocean," Jet Propulsion Laboratory, NASA, Caltech press release, January 10, 2000.

26. Fred Guterl, "Mission to Mercury," *Discover Magazine* 25, no. 4 (April 2004).

27. David Halliday and Robert Resnick, *Fundamentals of Physics*, 2nd ed. (New York: John Wiley & Sons, 1981), p. 271.

28. Carl Sagan, *Cosmos* (New York: Random House, 1980), p. 95.

29. Ibid., p. 107.

30. Percival Lowell, *Mars as the Abode of Life* (New York: MacMillan, 1910), p. 151.

31. Donald Savage and Guy Webster, "*Opportunity* Rover Finds Strong Evidence Meridiani Planum Was Wet," NASA press release 04-077, March 2, 2004.

32. Pasachoff, *Astronomy*, p. 216.

33. Ibid., p. 217.

34. Author witnessed *Pathfinder* landing during the Planetary Society's *Planetfest* event, Pasadena Convention Center, July 4, 1997.

35. Author witnessed *Spirit* landing during the Planetary Society's *Wild about Mars* event, Pasadena Convention Center, January 3, 2004—Pacific Standard Time.

36. Author witnessed *Opportunity* landing during Jet Propulsion Laboratory's Friends and Family event, Pasadena City College, January 24, 2004—Pacific Standard Time.

37. Kenneth Gatland, *Robot Explorers: The Pocket Encyclopedia of Spaceflight in Color* (London: Blandford Press, 1972), p. 217.

38. Freedman and Kaufmann, *Universe*, p. 262.

39. Pasachoff, *Astronomy*, p. 213.

40. Freedman and Kaufmann, *Universe*, p. 274.

41. Ibid., p. 277.

42. Yenne, *Atlas of the Solar System*, p. 107.

43. Freedman and Kaufmann, *Universe*, p. 285.

44. Magalhaes, "Galileo Probe Mission Science Summary."

45. Ibid.

46. Platt, "Galileo Findings Boost Idea of Other-Worldly Ocean."

47. Webster, "Solar System's Largest Moon Likely Has a Hidden Ocean."

48. Isbell and Platt, "Jupiter's Moon Callisto May Hide Salty Ocean."

49. Webster, "Solar System's Largest Moon Likely Has a Hidden Ocean."

50. Freedman and Kaufmann, *Universe*, p. 326.

51. Greeley and Batson, *NASA Atlas of the Solar System*, p. 242.

52. Yenne, *Atlas of the Solar System*, p. 128; Jet Propulsion Laboratory, "Cassini Saturn Arrival Mission Description," Jet Propulsion Laboratory, NASA, Caltech report, PD 699-100 Rev O, JPL D-5564 Rev O (June 2004), p. 10.

53. Freedman and Kaufmann, *Universe*, p. 351.

54. Information mathematically derived by author from data gathered from Pasachoff, *Astronomy*, pp. A-4, A-7, A-8.

55. Pasachoff, *Astronomy*, p. 309.

56. Ibid., 216.

57. Bennett et al., *The Solar System*, p. 243.

58. Ibid., p. 245.

59. Freedman and Kaufmann, *Universe*, p. 170.

ABOUT ASTEROIDS, COMETS, METEORS, AND OTHER SPACE JUNK

1. Jeffrey Bennett et al., *The Solar System: The Cosmic Perspective*, 3rd ed. (San Francisco: Addison Wesley, 2004), p. 369.

2. Bill Yenne, *Atlas of the Solar System* (New York: Bison Books, 1987), p. 179.

3. European Space Agency (ESA), "New Study Reveals Twice as Many Asteroids as Previously Believed," ESA Infrared Space Observatory, Science & Technology, http://sci.esa.int/ science-e/www/object/index.cfm?fobjectid=29762 (accessed April 5, 2002).

4. Minor Planet Center, "Lists and Plots: Minor Planets," Minor Planet Center, Smithsonian Astrophysical Observatory, International Astronomical Union, http://cfa-www.harvard.edu/ iau/lists/MPLists.html (accessed September 19, 2004).

5. Roger A. Freedman and William J. Kaufmann III, *Universe*, 6th ed. (New York: W. H. Freeman and Company, 2002), p. 366.

6. Southwest Research Institute, "Observations Reveal Curiosities on the Surface of Asteroid Ceres," *Southwest Research Institute News*, http://www.boulder.swri.edu/recent/ceres.html (accessed October 19, 2001).

7. National Space Science Data Center (NSSDC), "Asteroid Fact Sheet," NSSDC, NASA Goddard Space Flight Center, http:// nssdc.gsfc.nasa.gov/planetary/factsheet/asteroidfact.html (accessed September 1, 2004).

8. Tom Standage, *The Neptune File: A Story of Astronomical Rivalry and the Pioneers of Planet Hunting* (New York: Walker, 2000), p. 30.

9. John Davies, *Beyond Pluto: Exploring the Outer Limits of the Solar System* (Cambridge: Cambridge University Press, 2001), p. 4.

10. NSSDC, "Asteroid Fact Sheet."

11. William E. Burrows, *Exploring Space: Voyages in the Solar System and Beyond* (New York: Random House, 1990), p. 275.

12. Information mathematically derived by author from data gathered from Freedman and Kaufmann, *Universe*, pp. 262, 285.

13. Ronald Greeley and Raymond Batson, *NASA Atlas of the Solar System* (Cambridge: Cambridge University Press, 1997), p. 313.

14. Minor Planet Center, "Trojan Minor Planets," Smithsonian Astrophysical Observatory, International Astronomical Union, http://cfa-www.harvard.edu/iau/lists/Trojans.html (accessed September 19, 2004).

15. Freedman and Kaufmann, *Universe*, p. 370.

16. James Lawrence Powell, *Night Comes to the Cretaceous* (New York: W. H. Freeman, 1998), pp. 176–79.

17. Bennett et al., *The Solar System*, pp. 386–87.

18. Powell, *Night Comes to the Cretaceous*, p. 103.

19. Freedman and Kaufmann, *Universe*, p. 377.

20. Ibid.

21. Jay M. Pasachoff, *Astronomy: From the Earth to the Universe* (Belmont, CA: Thomson Learning, 2002), p. 335.

22. Davies, *Beyond Pluto*, p. 16.

23. Freedman and Kaufmann, *Universe*, p. 379.

24. Pasachoff, *Astronomy*, p. 310.

25. Davies, *Beyond Pluto*, p. 45.

26. Ibid., p. 19.

27. The *Bayeux Tapestry* is an eleventh-century tapestry depicting the Norman invasion of England in 1066, Bayeux, France.

28. Yenne, *Atlas of the Solar System*, p. 166.

29. Giotto Mission, "Halley Flyby: 13 March 1986," European Space Agency, http://sci.esa.int/science-e/www/object/index.cfm?fobjectid=31878 (accessed July 16, 2003).

30. Freedman and Kaufmann, *Universe*, p. 375.

31. NSSDC, "Comet C/1996 B2 Hyakutake," NASA Goddard Space Flight Center, http://nssdc.gsfc.nasa.gov/planetary/hyakutake.html (accessed December 5, 2000).

32. NSSDC, "Comet Hale-Bopp," NASA Goddard Space Flight Center, http://nssdc.gsfc.nasa.gov/planetary/halebopp.html (accessed March 6, 2000).

33. Pasachoff, *Astronomy*, p. 353.

34. Ibid.

35. Ibid.

36. Tony Phillips, "Meteor Storm Forecast," Science@NASA, http://science.nasa.gov/headlines/y2002/09oct_leonidsforecast.htm (accessed October 9, 2002).

37. Information derived from several sources, including: Freedman and Kaufmann, *Universe*, p. 383; Pasachoff, *Astronomy*, p. 357; Bennett et al., *The Solar System*, p. 386; Harvard-Smithsonian Center for Astrophysics, "Major Annual Meteor Showers," http://cfa-www.harvard.edu/cfa/ep/meteor/shower1.html (accessed March 27, 2001); and National Weather Service Forecast Office, "Meteors and Meteor Showers," National Oceanic and Atmospheric Administration, http://www.crh.noaa.gov/fsd/astro/meteor.htm (accessed June 22, 2004).

38. Minor Planet Center, "Unusual Minor Planets," Smithsonian Astrophysical Observatory, International Astronomical Union, http://cfa-www.harvard.edu/iau/lists/Unusual.html (accessed September 19, 2004).

39. Freedman and Kaufmann, *Universe*, p. 382.

40. David Bellingham, *An Introduction to Greek Mythology* (Secaucus, NJ: Chartwell Books, 1989), pp. 7, 81.

41. Davies, *Beyond Pluto*, p. 22.

42. Ibid., pp. 25–29.

43. Ibid., 32.

44. NSSDC, "Chiron Perihelion Campaign," NASA Goddard Space Flight Center, http://nssdc.gsfc.nasa.gov/planetary/chiron.html (accessed December 11, 2003).

ABOUT EARTH

1. Jay M. Pasachoff, *Astronomy: From the Earth to the Universe* (Belmont, CA: Thomson Learning, 2002), p. 685.

2. William L. Masterton, Emil J. Slowinski, and Conrad L. Stanitski, *Chemical Principles*, 5th ed. (Philadelphia: Saunders College Publishing), p. 360.

3. Pasachoff, *Astronomy*, pp. 146–47.

4. Roger A. Freedman and William J. Kaufmann III, *Universe*, 6th ed. (New York: W. H. Freeman and Company, 2002), p. 200.

ABOUT THE MOON

1. Philip Hiscock, "Once in a Blue Moon," *Sky & Telescope* (March 1999): 53.

2. Donald W. Olson, Richard Fienberg, and Roger Sinnott, "What's a Blue Moon?" *Sky & Telescope* (May 1999): 36–38.

3. Ibid.

4. Information calculated using *TheSky* astronomy software (Golden, CO: Software Bisque, 1998).

5. "Full Moon Names and Their Meanings," *Farmer's Almanac*, http://www.farmersalmanac.com/astronomy/fullmoonnames.html (accessed September 19, 2004).

6. Roger A. Freedman and William J. Kaufmann III, *Universe*, 6th ed. (New York: W. H. Freeman and Company, 2002), p. 217.

7. Jay M. Pasachoff, *Astronomy: From the Earth to the Universe* (Belmont, CA: Thomson Learning, 2002), p. 160.

8. Freedman and Kaufmann, *Universe*, p. 216.

9. Ibid., 207.

10. Neil Comins, *What If the Moon Didn't Exist?: Voyages to Earths That Might Have Been* (New York: HarperCollins, 1993), pp. 2–4.

11. Pasachoff, *Astronomy*, p. 161.

12. Fred Espenak, "Lunar Eclipse Page," NASA Goddard Space Flight Center, http://sunearth.gsfc.nasa.gov/eclipse/lunar.html (accessed September 9, 2004).

13. Pasachoff, *Astronomy*, p. 156.

14. National Space Science Data Center (NSSDC), "*Luna 3*," NASA Goddard Space Flight Center, http://nssdc.gsfc.nasa.gov/database/MasterCatalog?sc=1959-008A (accessed April 5, 2001).

ABOUT SPACECRAFT AND SPACE TRAVEL

1. Brad Wright, "First Space Shuttle Finds a New Home," CNN.com, Science & Space, http://www.cnn.com/2003/TECH/space/11/24/shuttle.enterprise/ (accessed November 24, 2003).

2. Kerry Mark Joels and Gregory P. Kennedy, *The Space Shuttle Operator's Manual* (New York: Ballantine Books, 1982), 1.7.

3. Joseph Yoon, "Launch Speeds and Earth Rotation," aero-

spaceweb.org, http://www.aerospaceweb.org/question/spacecraft/q0115b.shtml (accessed March 16, 2003).

4. David Halliday and Robert Resnick, *Fundamentals of Physics*, 2nd ed. (New York: John Wiley & Sons, 1981), p. A5.

5. Kennedy Space Center, "Orbiter Thermal Protection System," NASA Facts Online, KSC release no. 11-89, http://www-pao.ksc.nasa.gov/kscpao/nasafact/tps.htm (accessed February 1989).

6. Harold W. Gehman Jr. et al., *Columbia Accident Investigation Board: Report Volume 1* (Washington, DC: US Government Printing Office, August 2003), p. 34.

7. Ibid., pp. 65–73.

8. Ibid., p. 3.

9. William P. Rogers et al., *Report of the Presidential Commission of the Space Shuttle Challenger Accident*, vol. 1 (Washington DC: US Government Printing Office, June 1986), p. 19.

10. Ibid., pp. 19–21.

11. Ibid., p. iii.

12. David Baker, *History of Manned Space Flight* (New York: Crown Publishers, 1981), p. 535.

13. Ibid., p. 536.

14. Information derived from the following sources: Antonín Rükl, *Atlas of the Moon* (Waukesha, WI: Kalmbach Publishing, 1996), pp. 134, 142; Charles A. Whitney, *Whitney's Star Finder* (New York: Alfred A. Knopf, 1980), p. 63; and author's observing experience.

15. National Space Science Data Center (NSSDC), *"Luna 2,"* NASA Goddard Space Flight Center, http://nssdc.gsfc.nasa.gov/database/MasterCatalog?sc=1959-014A (accessed November 19, 2003).

16. Kenneth Gatland, *Robot Explorers: The Pocket Encyclopedia of Spaceflight in Color* (London: Blandford Press, 1972), p. 134.

17. Alan Shepard and Deke Slayton, *Moonshot: The Inside Story of America's Race to the Moon* (Atlanta: Turner Publishing, 1994), p. 226.

18. NSSDC, *"Zond 5,"* NASA Goddard Space Flight Center, http://nssdc.gsfc.nasa.gov/database/MasterCatalog?sc=1968-076A (accessed February 14, 2001).

19. NSSDC, *"Luna 24,"* NASA Goddard Space Flight Center, http://nssdc.gsfc.nasa.gov/database/MasterCatalog?sc=1976-081A (accessed March 23, 2004).

20. NSSDC, *"Venera 7,"* NASA Goddard Space Flight Center, http://nssdc.gsfc.nasa.gov/database/MasterCatalog?sc=1970-060A (accessed November 21, 2003).

21. Information derived from the following sources: Jay M. Pasachoff, *Astronomy: From the Earth to the Universe* (Belmont, CA: Thomson Learning, 2002), p. 196; and NSSDC, "Chronology of Venus Exploration," NASA Goddard Space Flight Center, http://nssdc.gsfc.nasa.gov/planetary/chronology_venus.html (accessed September 9, 2004).

22. Jet Propulsion Laboratory, "Voyager Interstellar Mission," NASA, Jet Propulsion Laboratory, http://voyager.jpl.nasa.gov/mission/weekly-reports/index.htm (accessed July 30, 2004).

23. Anatoly Zak, "Lunar Probes: The Overview of the Russian Launches to the Moon," http://www.russianspaceweb.com/spacecraft_planetary_lunar.html (accessed September 20, 2004).

24. Jet Propulsion Laboratory, "Genesis Mission Status Report," NASA, Jet Propulsion Laboratory news release, http://www.jpl.nasa.gov/news/news.cfm?release=2004-231 (accessed September 16, 2004).

25. Jet Propusion Laboratory, "Stardust: NASA's Comet Sample Return Mission," NASA, Jet Propulsion Laboratory, http://stardust.jpl.nasa.gov/ (accessed September 20, 2004).

26. Jet Propulsion Laboratory, "Deep Space Network," NASA, Jet Propulsion Laboratory, http://deepspace.jpl.nasa.gov/dsn/ (accessed September 13, 2004).

ABOUT STARS

1. Roger A. Freedman and William J. Kaufmann III, *Universe*, 6th ed. (New York: W. H. Freeman and Company, 2002), p. 33.

2. Ibid.

3. International Astronomical Union (IAU), "Naming Stars," http://www.iau.org/IAU/FAQ/starnames.html (accessed February 25, 2003).

4. IAU, "Welcome to the IAU World Wide Web Page," http://www.iau.org/ (accessed September 20, 2004).

5. IAU, "Naming Stars."

6. Charles A. Whitney, *Whitney's Star Finder* (New York: Alfred A. Knopf, 1980), p. 56.

7. Jay M. Pasachoff, *Astronomy: From the Earth to the Universe* (Belmont, CA: Thomson Learning, 2002), p. 448.

8. Hipparcos Space Astrometry Mission, "Hipparcos: Background Information," European Space Agency, http://astro.estec.esa.nl/Hipparcos/further_more.html#themission (accessed September 10, 1999).

9. Freedman and Kaufmann, *Universe*, p. 432.

10. Ibid., p. 437.

11. Ibid., p. 440.

12. Pasachoff, *Astronomy*, pp. 485–90.

13. Freedman and Kaufmann, *Universe*, p. 590.

14. Ian Ridpath, *Dictionary of Astronomy* (New York: Oxford University Press), p. 223.

15. Pasachoff, *Astronomy*, p. 704.

16. Robert Burnham Jr., *Burnham's Celestial Handbook: An Observer's Guide to the Universe beyond the Solar System*, 3 vols. (New York: Dover Publishing, 1978), 1:549.

17. Ibid., 2:916.

18. H. J. P. Arnold, P. Doherty, and P. Moore, *Photographic Atlas of the Stars* (Waukesha, WI: Kalmbach Publishing, 1997), p. 112.

19. Freedman and Kaufmann, *Universe*, p. 171; Pasachoff, *Astronomy*, p. 325.

20. Edmund Scientific Company, *Edmund MAG 5 Star Atlas* (Barrington, NJ: Edmund Scientific, 1980), p. 13.

21. Freedman and Kaufmann, *Universe*, p. 508.

22. Jeffrey Bennett et al., *Stars, Galaxies & Cosmology: The Cosmic Perspective*, 3rd ed. (San Francisco, CA: Addison Wesley, 2004), p. 580.

23. Freedman and Kaufmann, *Universe*, p. 616.

ABOUT BLACK HOLES

1. David Halliday and Robert Resnick, *Fundamentals of Physics*, 2nd ed. (New York: John Wiley & Sons, 1981), p. A5.

2. Jay M. Pasachoff, *Astronomy: From the Earth to the Universe* (Belmont, CA: Thomson Learning, 2002), p. 596.

3. Stuart L. Shapiro and Saul A. Teukolsky, *Black Holes, White Dwarfs, and Neutron Stars: The Physics of Compact Objects* (New York: John Wiley & Sons, 1983), p. 335.

4. Jeffrey Bennett et al., *Stars, Galaxies & Cosmology: The Cosmic Perspective*, 3rd ed. (San Francisco, CA: Addison Wesley, 2004), p. 583.

5. Stephen Hawking, *A Brief History of Time* (New York: Bantam Books, 1998), p. 109.

6. Ibid., p. 110.

7. Roger A. Freedman and William J. Kaufmann III, *Universe*, 6th ed. (New York: W. H. Freeman and Company, 2002), p. 588.

8. Pasachoff, *Astronomy*, p. 627.

9. Hawking, *Brief History of Time*, p. 101.

10. Freedman and Kaufmann, *Universe*, p. 552.

11. Bennett et al., *Stars, Galaxies & Cosmology*, pp. 586–87.

12. Freedman and Kaufmann, *Universe*, p. 555.

13. Hawking, *Brief History of Time*, p. 110.

14. Ibid., p. 112.

ABOUT GALAXIES

1. Roger A. Freedman and William J. Kaufmann III, *Universe*, 6th ed. (New York: W. H. Freeman and Company, 2002), p. 566.

2. Robert Burnham Jr., *Burnham's Celestial Handbook: An Observer's Guide to the Universe beyond the Solar System*, 3 vols. (New York: Dover Publishing, 1978), 1:131.

3. H. J. P. Arnold, P. Doherty, and P. Moore, *Photographic Atlas of the Stars* (Waukesha, WI: Kalmbach Publishing, 1997), p. 56.

4. Jay M. Pasachoff, *Astronomy: From the Earth to the Universe* (Belmont, CA: Thomson Learning, 2002), p. 627.

5. Ibid.

6. Agnès Villanueva, Jacqueline Mitton, and Philippe Chauvin, "Astronomers Find Nearest Galaxy," Université Louis Pasteur-Strasbourg, Observatoire de Strasbourg, http://astro.u-strasbg.fr/images_ri/canm-e.html (accessed November 4, 2003).

7. Daniel Schaerer and Roser Pelló, "VLT Smashes the Record of the Farthest Known Galaxy," European Southern Observatory, http://www.eso.org/outreach/press-rel/pr-2004/pr-04-04.html (accessed March 1, 2004).

8. Ibid.

9. Robert Roy Britt, "Most Distant Galaxy Hints at Dark Ages," Space.com, Science, http://www.space.com/scienceastronomy/distant_galaxy_040216.html (accessed February 16, 2004).

ABOUT TELESCOPES

1. Eugene Hecht and Alfred Zajac, *Optics* (Reading, MA: Addison-Wesley, 1974), p. 151.

2. Carl Sagan, *Cosmos* (New York: Random House, 1980), pp. 142, 145.

3. Jay M. Pasachoff, *Astronomy: From the Earth to the Universe* (Belmont, CA: Thomson Learning, 2002), p. 40.

4. Ibid.

5. Jeffrey Bennett et al., *Stars, Galaxies & Cosmology: The Cosmic Perspective*, 3rd ed. (San Francisco, CA: Addison Wesley, 2004), p. 22.

6. W. M. Keck Observatory, "The Observatory," http://www2.keck.hawaii.edu/geninfo/about.html (accessed September 22, 2004).

7. "Hobby-Eberly Telescope," University of Texas, Austin, http://www.as.utexas.edu/mcdonald/het/het.html (accessed September 9, 2004); Very Large Binocular Telescope Observatory, "Telescope," http://medusa.as.arizona.edu/lbto/telescop.html (accessed September 22, 2004); European Southern Observatory, "VLT Information," http://www.eso.org/outreach/ut1fl/ (accessed September 10, 2004); National Astronomical Observatory of Japan, "What Is Subaru?" Subaru Telescope, http://www.naoj.org/Introduction/outline.html (accessed September 22, 2004); Gemini Observatory, "About the Gemini Observatory," Association of Universities for Research in Astronomy and National Science Foundation, http://www.gemini.edu/index.php?option=content&task=view&id=9&Itemid=2 (accessed September 22, 2004).

8. Robert Tindol and Richard Ellis, "Gordon and Betty Moore Foundation Awards $17.5 Million for Thirty-Meter Telescope Plans," California Institute of Technology press release, http://pr.caltech.edu/media/Press_Releases/PR12441.html (accessed October 16, 2003).

9. Hubble Space Telescope: New Views of the Universe, "The Telescope," NASA, http://hubblesite.org/discoveries/hstexhibit/telescope/about.shtml (accessed September 22, 2004).

A BRIEF HISTORY OF LUNAR AND PLANETARY EXPLORATION

Listed alphabetically by author/editor or title (if no named author available).

Baker, David. *The History of Manned Space Flight*. New York: Crown Publishers, 1981.

Christy, Robert. "Zarya: A Source of Information on Soviet and Russian Spaceflight." http://www.zarya.info/ (accessed September 22, 2004).

Fimmel, Richard O., William Swindell, and Eric Burgess. *Pioneer Odyssey*. Washington, DC: US Government Printing Office, 1977.

Gatland, Kenneth. *Robot Explorers: The Pocket Encyclopedia of Spaceflight in Color*. London: Blandford Press, 1972.

Godwin, Robert, ed. *Mars, The NASA Mission Reports: Volume 2*. Washington, DC: US Government Printing Office, 2004.

Jet Propulsion Laboratory. "Cassini-Huygens Mission to Saturn." NASA, Caltech. http://saturn.jpl.nasa.gov/home/index.cfm (accessed September 22, 2004).

————. "Mars Exploration Rover Mission." NASA, Caltech. http://marsrovers.jpl.nasa.gov/home/ (accessed September 24, 2004).

————. "Voyager Interstellar Mission." NASA, Caltech. http://www.astronautix.com/craft/index.htm (accessed September 22, 2004).

Kohlhase, Charles, ed. *The Voyager Neptune Travel Guide*. Washington, DC: US Government Printing Office, 1989.

Murray, Bruce, and Eric Burgess. *Flight to Mercury*. New York: Columbia University Press, 1977.

NASA. "Pioneer Project Home Page." http://spaceprojects.arc.nasa.gov/Space_Projects/pioneer/PNhome.html.

NASA Pocket Statistics. Washington, DC: US Government Printing Office, 1996.

NSSDC. "Chronology of Lunar and Planetary Exploration." NASA Goddard Space Flight Center. http://nssdc.gsfc.nasa.gov/planetary/ chrono.html (accessed September 1, 2004).

Spitzer, Cary R., ed. *Viking Orbiter Views of Mars*. Washington, DC: US Government Printing Office, 1980.

Wade, Mark. "Encyclopedia Astronautica: Spacecraft." http://www.astronautix.com/craft/index.htm (accessed September 22, 2004).

Yenne, Bill. *Atlas of the Solar System*. New York: Bison Books, 1987.

Young, Carolynn, ed. *The Magellan Venus Explorer's Guide*. Pasadena, CA: NASA Jet Propulsion Laboratory, 1990.

Zak, Anatoly. "News & History of Astronautics in the former USSR: Spacecraft." http://www.russianspaceweb.com/spacecraft.html (accessed September 22, 2004).

INDEX